REVIS

GEARS & GEAR CUTTING

FOR HOME MACHINISTS

IVAN LAW

Revised and updated by **George Bulliss**,
editor of *The Home Shop Machinist* magazine

Parts of this book were updated for today's American reader with regard to new techniques and tools. These updates were graciously provided by George Bulliss of *The Home Shop Machinist*, *Machinist's Workshop*, and *Digital Machinist* magazines.

First published in the United Kingdom by Argus Books
© Special Interest Model Books Ltd, 2003
First published in North America in 2018 by Fox Chapel Publishing.

Cover image courtesy Shutterstock: Florin Patrunjel

ISBN 978-1-56523-917-3

Library of Congress Cataloging-in-Publication Data

Names: Law, Ivan R., author.
Title: Gears & gear cutting for home machinists / Ivan R. Law.
Other titles: Gears and gear cutting for home machinists
Description: Mount Joy : Fox Chapel Publishing, [2018] | Includes index.
Identifiers: LCCN 2017058617 | ISBN 9781565239173 (pbk.)
Subjects: LCSH: Gearing. | Gear-cutting machines.
Classification: LCC TJ184 .L39 2018 | DDC 621.8/33--dc23
LC record available at https://lccn.loc.gov/2017058617

Printed in Singapore
Third printing

CONTENTS

PREFACE 4

ACKNOWLEDGMENTS 5

CHAPTER 1:
BASICS 6

CHAPTER 2:
TOOTH FORM 12

CHAPTER 3:
GEAR TOOTH SIZES 27

CHAPTER 4:
RACK AND PINION GEARS 30

CHAPTER 5:
BEVEL WHEELS 35

CHAPTER 6:
WORM GEARS 45

CHAPTER 7:
DEFINITIONS AND FORMULAS. 53

CHAPTER 8:
DIVIDING HEADS 57

CHAPTER 9:
CUTTING SPUR GEARS 62

CHAPTER 10:
CUTTING WORMS AND WORM WHEELS 83

CHAPTER 11:
CUTTING BEVEL GEARS 102

CHAPTER 12:
MAKING GEAR CUTTERS 110

APPENDIX 133

INDEX 134

PREFACE

Gearing, if studied deeply, can be—in fact is—a complicated and highly technical subject on which many books have been written, often by people who have spent their entire academic lives studying the many problems involved. This is not one of those books; this is not for the technical student who is on his way to a PhD in engineering but for the average man who, in his back garden workshop, enjoys model-making or just tinkering about with mechanisms and may wish to use a pair of gears in some project. Rather than modifying or impairing his design by trying to work-in commercial gears, he may wish to produce his own.

But often gears do represent a threshold over which the amateur, through lack of information, is hesitant to step and I have written this book with the express purpose of encouraging these people to "have a go" and to show them that the design and manufacture of a pair of gears is well within their capacity. No special knowledge is needed, just plain commonsense, which is never lacking with model engineers. I have endeavored wherever possible to use plain, simple and non-technical language and have kept the level of mathematics down to the simplest form. Indeed, any reader whose mathematical education extended no further than reciting his multiplication tables will, if he follows the methods outlined, have no difficulty in satisfactorily solving the problems that arise in designing and producing a pair of spur gears.

This does not mean that corners have been cut, nor that the information given is not theoretically correct. Neither does it mean that the gears produced as a result of following the information given in this book will be in any way inferior to commercially produced ones. What has been done is to approach the question of gears, not from the scientific or technical point of view, but from a purely practical standpoint.

It is hoped that not only the potential constructor of his own gears but also any reader who merely wishes to obtain a basic understanding of gearing will find this book of considerable interest and assistance.

ACKNOWLEDGMENTS

The author would like to thank two of his long-standing friends for the help he has received from them. Firstly, Mr. C. Tissiman, who volunteered to read the manuscript prior to publication. Mr. Tissiman has, over the years, cut many gears for small mechanisms and has the reputation of being able to repair and completely restore to their original condition musical boxes and all other mechanical musical devices whose gearing often defies all conventional principles! Secondly, Professor D. H. Chaddock, whose considerable knowledge on engineering matters is only overshadowed by his keen and eager willingness to help all who seek his counsel.

CHAPTER 1
BASICS

When faced with any problem the greatest step forward in finding a solution is to be able to fully understand just what the problem is. Once the problem is clearly understood then the first—and possibly the most important—hurdle has been overcome! Gears are no exception and amateur engineers who either try to ignore gears and pretend they are non-existent, or who get into trouble when producing them, have usually not begun at the beginning and asked themselves the question—"What are gears?"

Gears are used to transmit motion, and therefore power, between one shaft and another. To help to understand the principle involved it is advantageous to ignore the teeth on the gears and look upon them as discs with the outer surfaces in contact. For example, looking at Fig. 1, here we have two shafts A and B and we wish to transmit motion from

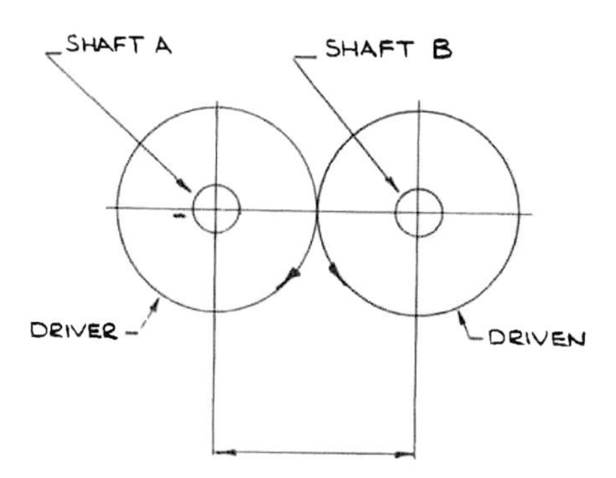

Fig. 1. *Center distance depends upon the disc diameters*

$$\frac{\text{SPEED OF SHAFT A.}}{\text{SPEED OF SHAFT B.}} = \frac{\text{DIA OF DISC B}}{\text{DIA OF DISC A}}$$

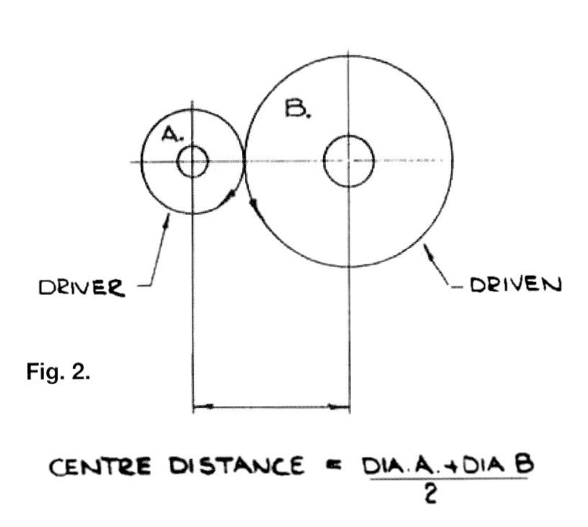

Fig. 2.

shaft A to shaft B. Two discs of equal diameter are used, one secured to each of the shafts. The diameters are such that the discs make contact and if we assume that no slip takes place between them, then as disc A rotates on its shaft it will in turn rotate disc B and hence shaft B. This is a simple diagram but quite a lot can be learned from it, the first thing being that the rotation of the two shafts will be in opposite directions.

As the drawing shows, the driver shaft A is turning clockwise and shaft B, which is called the driven shaft, must then rotate counterclockwise. If the design requirement is that both shafts should rotate in the same direction then this cannot be accomplished by using a pair of discs. The next thing we learn is that the diameter of the discs is determined by the center distance of the two shafts. In this case, since both discs are of equal diameter then the diameter of these discs will be the same as the center distance between the two shafts. Thirdly, as the circumference of both discs is the same length then the speed of rotation will also be similar.

Finally, since no slip between the two discs is taking place, then whatever the movement characteristics of shaft A happen to be, the same characteristics will be passed on to shaft B, albeit in the opposite direction. Or, to put it another way, the velocity between the two shafts will be constant. This last finding is of the utmost importance as the whole idea behind gear tooth design is to try to maintain this constant velocity. Even with all the advantages of modern technology friction drive using discs is often used if it is imperative to maintain a perfect velocity.

As a result of design requirements it may be necessary to have the two shafts A and B rotating at different speeds. Reference to Fig. 2 shows how this is achieved. If disc A is made smaller than disc B then the circumference of disc A will be less than that of disc B and therefore one complete revolution of shaft A will not result in one complete revolution of shaft B. It will be readily understood that the actual relationship between the speeds of the two shafts will be

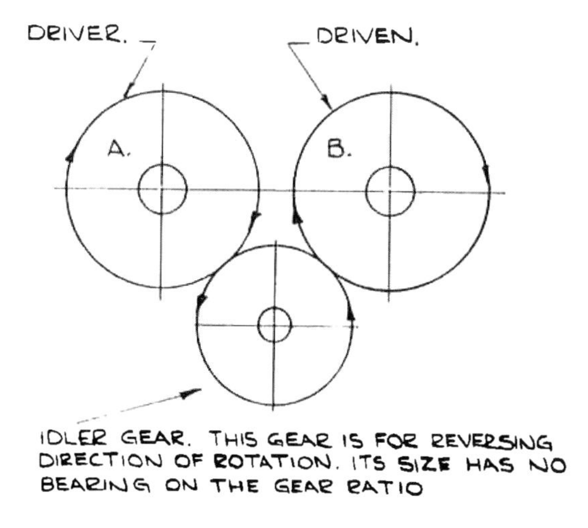

Fig. 3. *Introducing an idler enables both driven and driver to rotate in the same direction*

directly proportional to the circumferential distance of the two discs. If the distance around disc A is only one-half of the distance around disc B, then disc A will have to make two complete revolutions for one complete revolution of disc B—and so on.

When calculating circumferences of the discs the constant pi (π) has to be used and since this is, to say the least, a most unfortunate number as far as round figures are concerned it follows that the resultant figure of the circumferences will also be an awkward number. Fortunately, since pi is constant to both discs and therefore appears on both sides of the speed equation, it cancels itself out and so we can say that the speed of the discs is proportional to the diameter of the discs. If, therefore, we require a speed reduction of three, then the diameter of the driven disc B must be three times greater than the diameter of the driver disc A.

Referring again to Fig. 2, it will be seen that the distance between the two shafts is equal to the combined radii of the two discs, or to put it another way, Center Distance = ½ (dia. A + dia. B) and this equation is correct for any two touching discs. When two dissimilar diameters are used the small disc is usually referred to as the "pinion," while the larger disc is called the "wheel."

So far we have only considered two touching discs, but there may be any number and where more than two are used then the arrangement is called a "train." It was said earlier that if the two shafts A and B are to rotate in the same direction then a pair of discs could not be used; this is correct but a third disc could be introduced into the train whose only function would be to reverse the direction of rotation. This third disc is called an "idler," see Fig. 3. The size and position of the idler has no bearing at all on the speed

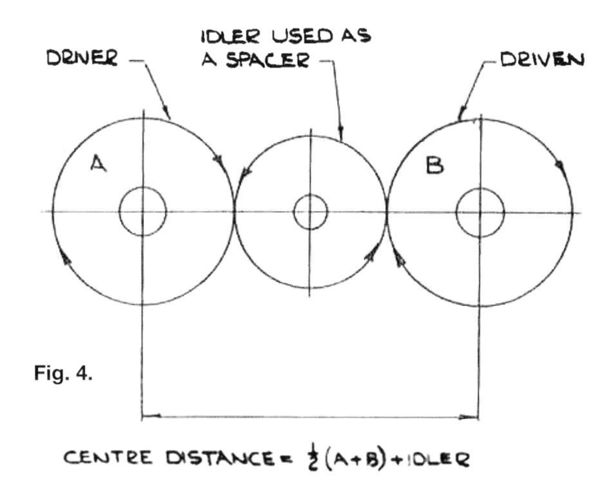

Fig. 4.

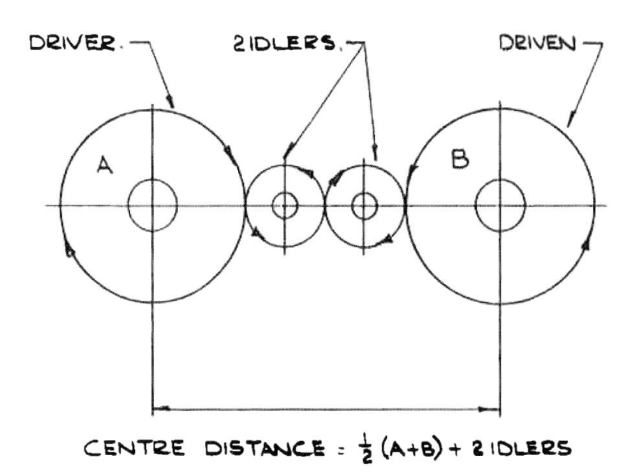

Fig. 5. *Two idlers used as spacers: driver and driven rotate in opposite directions*

or "gear" ratio between the original two discs. The relative speeds of A and B are still dependent on their respective diameters.

Usually, for design reasons, the idler disc is smaller than the other discs but it need not be so. It may be that the two shafts A and B are too far apart to allow two discs of sufficient diameter to be employed and so an idler disc is used to bridge the gap as shown in Fig. 4. Again, the size of the idler has no bearing on the gear ratio. It has, of course, reversed the direction of the driven gear B and this is illustrated in Figs. 3 and 4.

Should it be a design requirement that the driven gear has to rotate in the opposite direction to the driver gear, as in Figs. 1 and 2, then a second idler can be introduced into the train, as in Fig. 5. The two idlers take up the space and reverse the rotation of the driven disc B. All four discs are shown in one straight line but if the space available necessitates the use of idler wheels smaller than it is considered prudent to use, then the centers of the idler wheels could be staggered as shown in Fig. 6. There may be any number of wheels in a train but if all the wheels are connected to each other in a way similar to the diagrams shown, then it is only the diameters of the driver disc A and the driver disc B that are considered when determining the gear ratio of the train. The speed of the idlers will be dependent on their respective diameters but whatever these are it will have no effect on the final speed of the driven disc B.

There are many permutations in which a simple gear train can be arranged and in all the examples shown so far the end discs have been the driver and the driven discs but this need not be so. In Fig. 7 we have a situation where three driven wheels of the same diameter need to be in a straight line—discs B, C and E. The power source, or driver disc A, has been placed between the discs B and C and so drives both of them directly. Consequently both B and C rotate in the same direction; the third driven disc E has to be driven by means of an idler—disc D—which is serving a dual purpose of bridging the gap between discs C and E and also acting as a direction reverser in order to ensure that the disc E rotates in the same direction as B and C. This arrangement could be stretched out to extend the range of driven wheels, such as in a roller conveyor, but in every case the speed of rotation of each driven disc will be determined by the diameter of the driven disc in question and the diameter of the driving disc A. The size and speed of any other disc need not be considered.

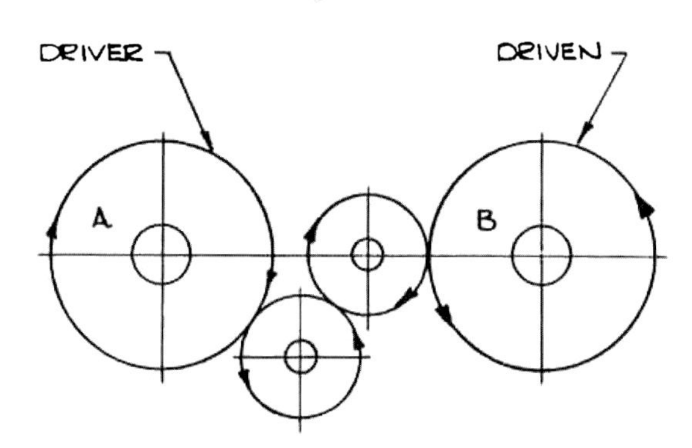

Fig. 6. *Staggered idlers reduce center distance of driver and driven discs*

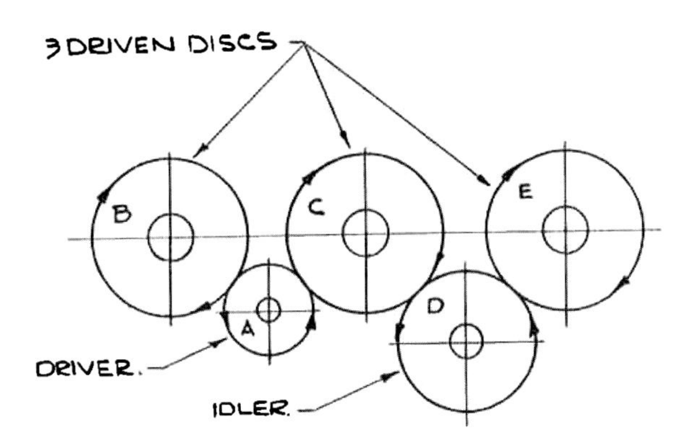

Fig. 7. *Gear train showing one driver, three driven, and one idler discs*

Up to now we have considered only rotating discs—not gears with teeth in them—and this is the correct path to follow as when designing and setting-out a gear drive or train they are looked upon as discs and all the setting-out and calculations are based upon the disc, not the gear!

Gearing, like any other subject, has its own terminology and the correct title for the disc is the "pitch circle diameter" often referred to as the PCD.

CHAPTER 2
TOOTH FORM

Up to now we have only considered discs rolling on each others surface and driving one another by the frictional forces between the two discs. If we refer back to Fig. 1 we can see that the driven wheel B will resist the moving action of the driving wheel, this resistance being a combination of the friction forces in the bearing and the useful work load we wish to obtain from the driven disc. As soon as the resistance to movement becomes greater than the friction grip between the surfaces of the two discs, slip will occur and the drive will fail. To overcome this problem teeth are added to the discs, which ensures positive engagement between the components and no loss of motion: the resulting discs are referred to as spur gears or spur wheels.

However, the solution of one problem often brings with it many more problems and putting teeth on the discs is certainly no exception. Referring to Fig. 8 we see here that a "lump," or projection, has been added to the wheel A. So, as the wheel rotates a groove or slot must be made in wheel B, otherwise the two wheels would jam together and movement would cease. When the projection enters the slot, the projection will act as a driving dog and slip cannot then take place; however, as soon as the projection disengages the slot,

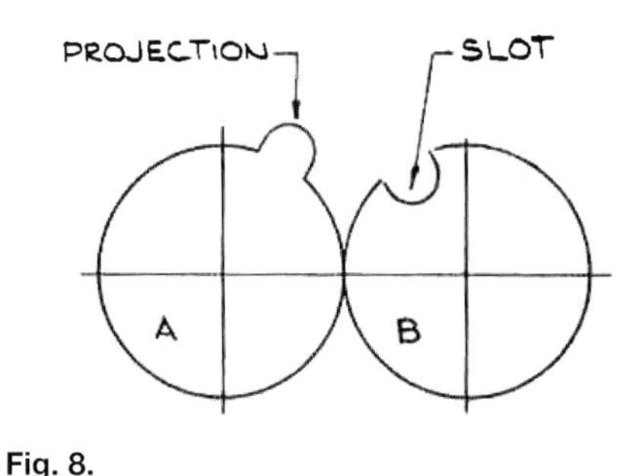

Fig. 8.

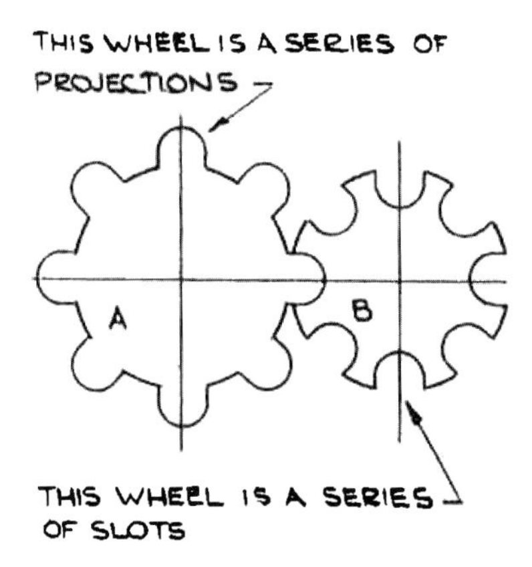

Fig. 9.

slip once more becomes possible. It is obvious therefore that a series of projections and slots must be provided so that at any point around the wheel there is always a projection engaged in a slot. This arrangement is shown at Fig. 9.

It can be seen that the two wheels are now dissimilar, one being a disc with projections around it while the other is a disc with grooves all the way around. There are spaces between each projection and slot which is the surface of the original disc and these spaces can be utilized by introducing an additional range of slots and projections, but in the case of this second set the projections are positioned between the slots on wheel B, and the slots are positioned between the projections on wheel A, the arrangement being that every projection is followed by a slot and every slot is followed by a projection. Each slot and projection combination is referred to as a "tooth," see Fig. 10. In arranging the series of teeth an unfortunate thing has happened in that we have completely lost the original disc. None of it remains and yet it has been the basis for all our setting-out and calculations. This is just one peculiar aspect of gears—the datum is the pitch circle diameter and yet it does not exist on the finished item. It cannot be seen and it cannot be touched so it cannot be utilized for direct measurements either from it or to it, but we must always consider a gear as a rolling PCD.

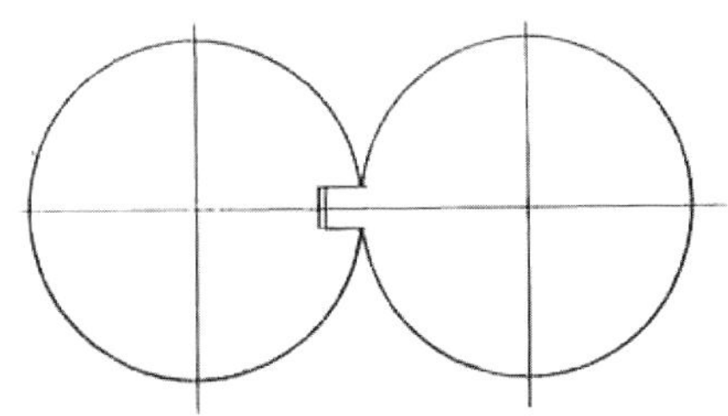

Fig. 11. *Rectangular teeth would lock the wheels together*

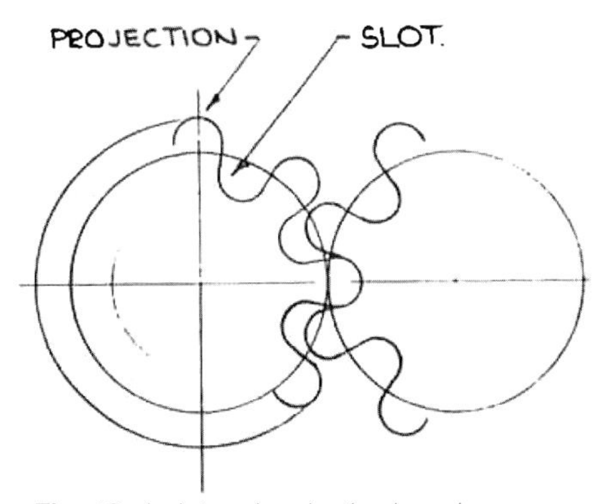

Fig. 10. *A slot and projection have become a tooth*

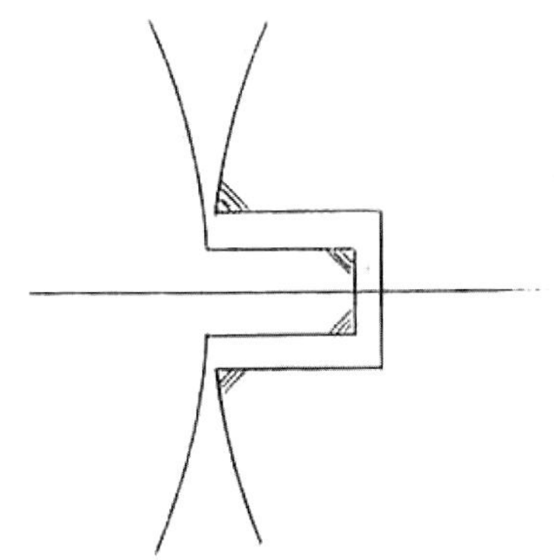

Fig. 12. *Enlarged view of rectangular "tooth" with clearance to allow rotation to take place*

Having established that each tooth now consists of two parts, the projection above the PCD and the groove or slot below it, it is not practical to make the projections and grooves any old shape. For example, supposing they were made rectangular, see Fig. 11, what has been done here is to key both wheels together and so rotation cannot take place at all! Rotation could become possible if a working clearance around the tooth were given as shown at Fig. 12, but an examination of the action is not encouraging. The corners of the tooth and the corners of the slot—areas shaded on the drawing—would in turn rub their way down the side of the slot and tooth as rotation took place. The end result would be excessive friction causing rapid wear and quickly leading to a complete failure of the teeth, and the noise such a pair of gears would generate would be painful on the ears, to say the least! However, the biggest disadvantage to this arrangement would be the complete loss of constant velocity between the two rolling PCDs. The driver wheel may turn with a constant and even velocity but the antics of the driven wheel would be difficult to visualize. It would rotate with a succession of stops, starts, accelerations, decelerations and judders, and the whole arrangement would, of course, be totally unacceptable.

This extreme example was taken to emphasize that in order to maintain the constant velocity obtained by the rolling PCD, the shape of the gear teeth is vitally important. A shape that simply allows rotation to take place will not necessarily be acceptable, it must also be a specific shape—a shape that allows the gears to rotate as though they were still discs.

There are two geometric curves that can be employed to give us the conditions we require. One curve is based on a shape called the cycloid and the second is based on the involute curve. Gears can be, and are, made to both standards but gears made to one standard must not be meshed with gears based on the other curve.

In the past the cycloidal type of gear was very much in favor while the involute stood in the wings. Nowadays, however, general engineering favors the involute gears. There are reasons for this change, the main one being that the involute curve can be easily generated and is therefore suitable for modern production methods. It could be argued that since the cycloidal gear is not often used in modern engineering this book should ignore it and concentrate on the popular involute gear but model engineers, who often look into the past for prototypes to model, will most likely become involved with the gearing based on the cycloid. Whether they substitute the modern involute gear in their recreation is, naturally, their choice but it must be an advantage to understand the basic principles involved in both types. That is why it is the intention of the author to give sufficient information for the reader to produce gear trains of both cycloidal and involute form; however, when it comes to cutting gears concentration will be on the involute.

THE CYCLOID CURVE

The definition of a cycloidal curve is as follows:

A cycloid is the curve which is described by a point fixed at the circumference of a circle when that circle is rolled in contact with a straight line. This may be better understood

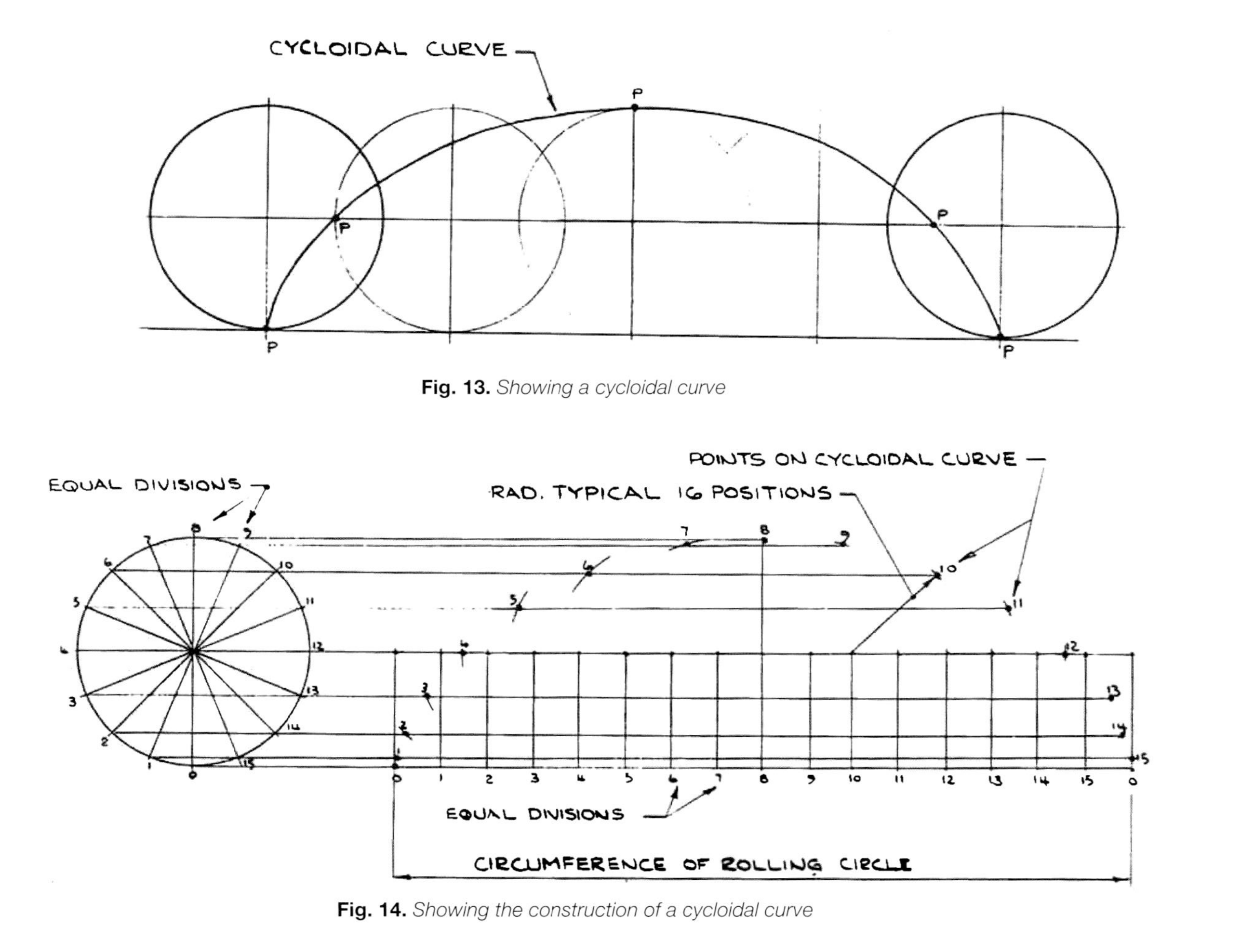

Fig. 13. *Showing a cycloidal curve*

Fig. 14. *Showing the construction of a cycloidal curve*

by referring to Fig. 13. Think of the circle drawn here as a locomotive wheel and the straight line as the track in front of that wheel. In position 1 the point of the wheel is in contact with the track; in the final position it is again on the track but the wheel has made one complete revolution. Intermediate positions of the wheel have been plotted to show the progress of the point. The resultant curve described by the point during the one complete turn of the wheel is the cycloidal curve.

It is not a difficult curve to draw and Fig. 14 shows how this can be done. A circle is drawn on the line and this circle is then divided into any number of parts—16 are shown in the example but the more divisions used the easier it will be to obtain a good curve. A length equal to the circumference of the circle is marked onto the base line and this line is then also divided into the same number of equal divisions as was chosen for the circle. These divisions are then projected upwards and onto the center line and so fix the centers of the circle at 16 different positions. Horizontal lines are then drawn from the 16 points around the original circle and where these lines intersect their relative circles they fix points on the cycloidal curve. In the diagram Fig. 14 point 10 has been chosen to illustrate the method. It can be seen that it is not necessary to draw the complete circle to fix the point, an arc crossing the horizontal line is all that is needed. A smooth curve is then drawn passing through all 16 points. It can be seen that the curve is symmetrical about the vertical center line.

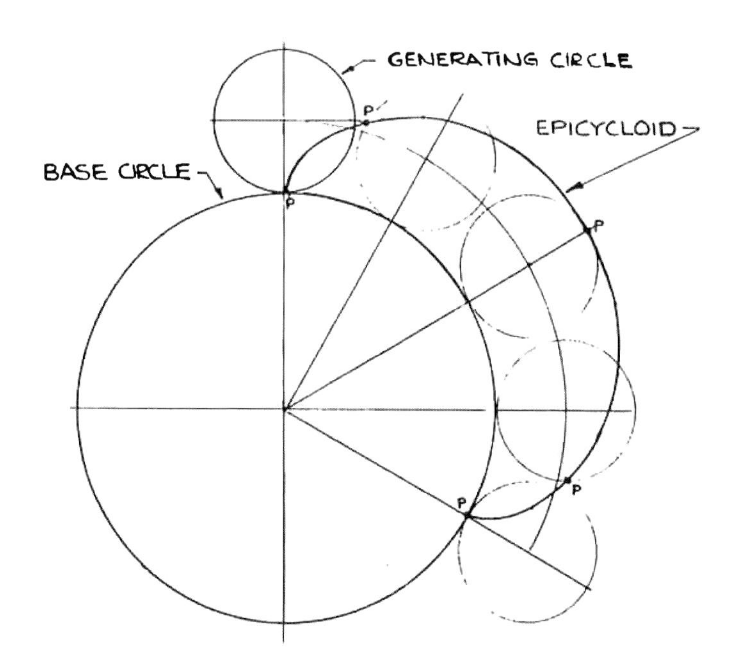

Fig. 15. *The epicycloidal curve*

If, however, instead of rolling the circle along a straight line it is rolled around the circumference of another circle, the resulting curve traced by a point on the rolling circle will produce a curve known as an epicycloid, see Fig. 15. The construction of the curve is basically similar to the cycloid except that the length of the circumference of the rolling circle is measured around the circumference of the base circle, the intermediate points being obtained by division as before. Another difference in construction is that the projections from the point on the rolling circle are not straight lines but are arcs taken from the center of the base circle, which can be swung around with the aid of a pair of compasses. This epicycloidal curve is important as the initial part of this curve is the correct shape for the gear tooth outside the PCD.

Going one step further, instead of rolling the generating circle, as it is called, outside the base circle it can be rolled inside it when another curve will be found. The construction of the new curve is basically similar to the epicycloid, see Fig. 16. This new curve is called the hypocycloid and represents the shape of a gear tooth below the PCD.

It can now be seen that the shape of a gear tooth of cycloidal form is a combination of two distinct curves joined together at the pitch circle diameter of the gear. Fig. 17 shows the shape of a typical gear tooth; it also shows that the surface above the PCD is usually called the "face" while the surface below the PCD is referred to as the "flank." The whole of the tooth above the PCD is the addendum while the whole of the tooth below the PCD is the dedendum.

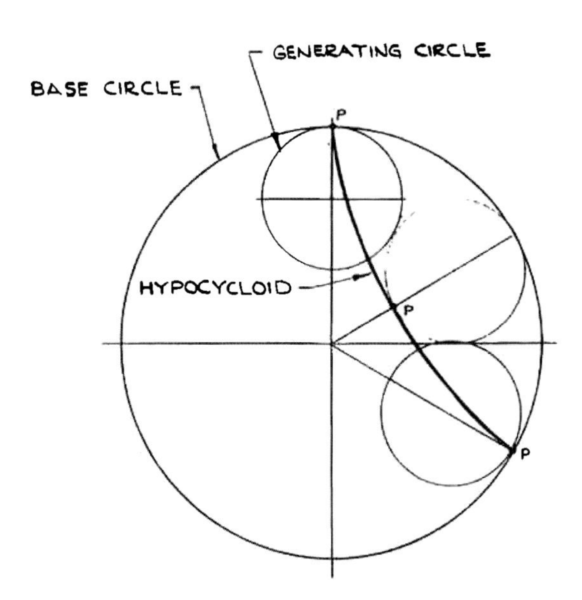

Fig. 16. *The hypocycloidal curve*

By studying the way in which the tooth shapes are determined it will be clearly seen that the shape of a gear tooth on any PCD can be influenced by the size of the rolling or generating circle. The size of the generating circle may vary but not in any two gears that are to mesh together. All gears based on the same rolling or generating circle will run together correctly and maintain constant velocity but gears having different generating circles must not be used in the same train. It is possible, but not usual when considering a pair of gears that only mesh with each other, to have one generating circle for the flanks of the driver teeth and faces of the driven teeth, and another generating circle for the flanks of the driven teeth and faces of the driver teeth.

When setting out a gear train from scratch, the size of the generating circle must be determined and often the size chosen is half the PCD of the smallest pinion. This ratio is interesting because the hypocycloid generated by a rolling circle whose diameter is half the PCD is a straight line. It is easy to verify this for yourself by simply drawing one following the instructions already given. This means, of course, that the flanks of the teeth on such a gear are straight-sided and radial, see Fig. 18. Even though the flanks are straight lines the gears will still roll together correctly and maintain constant velocity.

ADDENDUM
FACE
FLANK
EPICYCLOID.
P.C.D.
HYPOCYCLOID
DEDENDUM

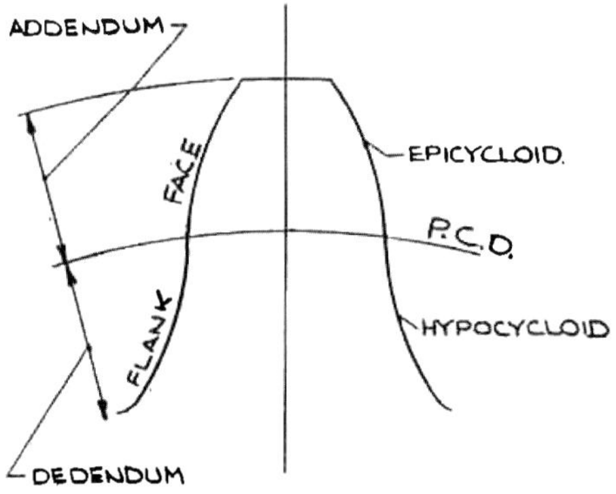

Fig. 17. *A typical cycloidal tooth*

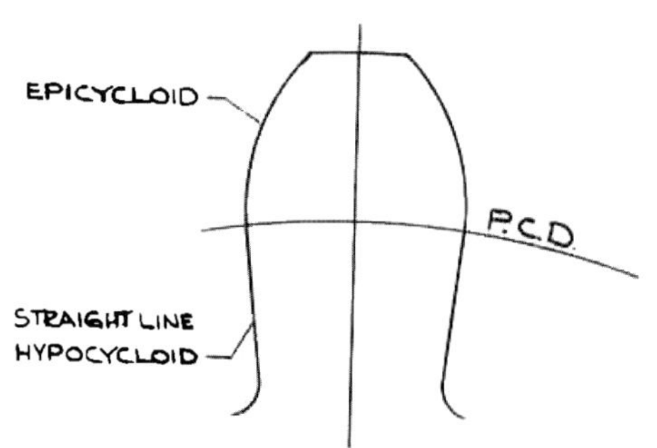

Fig. 18. *Showing the tooth profile using a generating circle whose diameter is a half of the PCD*

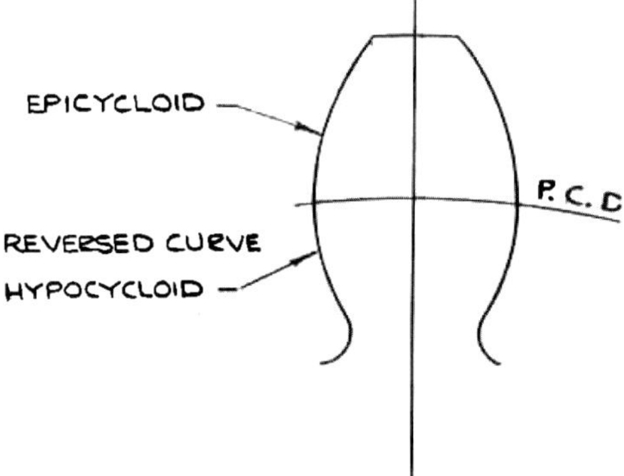

Fig. 19. *Showing the effect on tooth when the diameter of the generating circle is greater than a half of the PCD*

It is possible to increase the size of the generating circle further and still produce correctly formed gears; however, doing this produces a flank with a reversed curve. This undercuts the root of the tooth and so produces a weak form of tooth. This type of gear is only used for lightly loaded lowly-stressed duties, see Fig. 19.

There is no ideal ratio between the PCD and the generating circle, otherwise it would always be used and become a standard. One eminent gear designer suggests that 2.22 times the circular pitch is a good base from which to start.

TOOTH CONTACT

If one gear has to pass its power onto another gear then it is obvious that the teeth of the two gears must make contact with each other while in mesh. When discussing the impractical rectangular tooth shape earlier it was pointed out that a great amount of friction was caused by the "teeth" rubbing together. Friction is the main enemy of the engineer as not only is it responsible for power losses but it also generates heat and causes rapid wear. Gears are no exception and every effort must be made to reduce the rubbing action between the engaged teeth. This is another reason why gear tooth shape is important. Ideally the teeth should not rub against each other but merely roll over one another. If the rubbing action is removed then so is most of the friction.

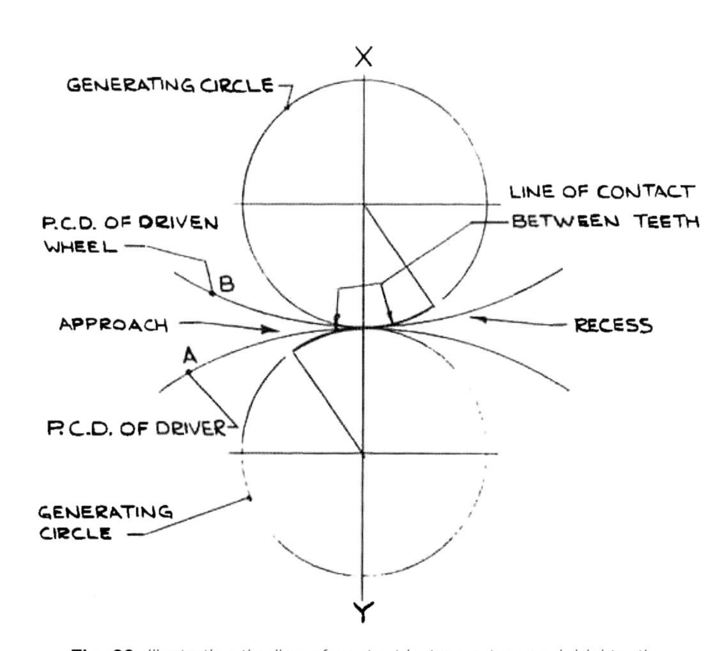

Fig. 20. *Illustrating the line of contact between two cycloidal teeth*

Fortunately, the shape that gives us "constant velocity" also gives us the basic rolling action. As the teeth come into mesh and out again, contact has to be made and this contact is not haphazard but follows a definite path, or line of action. This is illustrated in Fig. 20, which shows that the line of contact follows the shape of the generating circle of the driver gear A, but only up to the center line passing between the two gears—line XY. Here the line of action leaves the generating circle of wheel A to follow the path of the generating circle of wheel B until disengagement is complete. The line of action up to the center line is termed the "arc of approach" while the line of contact after the center line is called the "arc of recess." As in the case of the shape of the gear tooth, the line of action is a combination of two curves that join together at the PCDs.

LANTERN PINIONS

It is an engineering fact that the friction that is generated during the engagement of the teeth is far greater than the friction arising during disengagement. Referring to Fig. 20 the friction along the line of contact left of the line XY is greater, thereby causing more wear and power loss, than the friction that occurs after or to the right of the line XY. This is only correct so long as A is the driver; should A become the driven gear then the roles will be reversed and the high friction side would be to the right, which would then be the new approach side. Referring back to the "arc of approach" the actual parts of the teeth that would be in contact are the flanks of the driver gear A and the faces of the driven gear B, that is the "root" of the teeth on the driver and the "tops" of the teeth on the driven. After the center line is passed, the point of contact moves up the teeth so that it is now the faces of the driver that make contact with flanks of the driven gear.

There are occasions when the elimination of as much friction as is possible becomes paramount over other considerations and such a situation arises in the gear trains of clocks where little power needs to be transmitted, the clockmaker being interested mainly in motion. To conserve power by reducing friction is the prime consideration and since the high friction point of the arc of contact is the approach then this is where attention must be concentrated. The obvious solution is to eliminate the whole of the contact before the center line is reached, which means completely removing all the flanks off the teeth of the driver. This drastic action renders the faces of the driven wheel redundant and so they can go also. What remains is a driver wheel with no teeth below the pitch circle and a driven wheel with no teeth above the PCD. It was seen that altering the size of the generating circle had a considerable effect upon the cycloidal curves. If the size of the generating circle were to be made the same diameter as the PCD of the driven wheel which will be regarded as a pinion, then the shape of the resulting hypercycloid on the driven wheel becomes a point. The form of the wheel and pinion is now as shown in Fig. 21.

Obviously the arrangement is impractical and therefore some modifications must be made to the points on the pinion so that some useful advantage can be gained from this theoretical situation. Euclid defines a point as having position but no magnitude. In order to obtain magnitude a circle of material can be added completely around the points on the pinion thus making them into pins, the center of the pins thus becoming the original points.

The behavior of the pins will theoretically be similar to that of the original points and for all practical purposes the pins have become hypercycloids. Enlarging the points into pins will necessitate modification to the wheel in order to make room for the pins to engage with the teeth of the wheel. Moving the wheel centers farther apart to give the clearance will not do as this would mean that the PCD of the two gears would no longer be in contact and this condition must be at all times be preserved.

The solution to the problem is to remove or cut away part of the teeth on the wheel so as to make room for the pins on the pinion. If the portion cut away from each tooth is one-half of the pin's diameter and with the cutting line parallel to the original outline of the tooth then

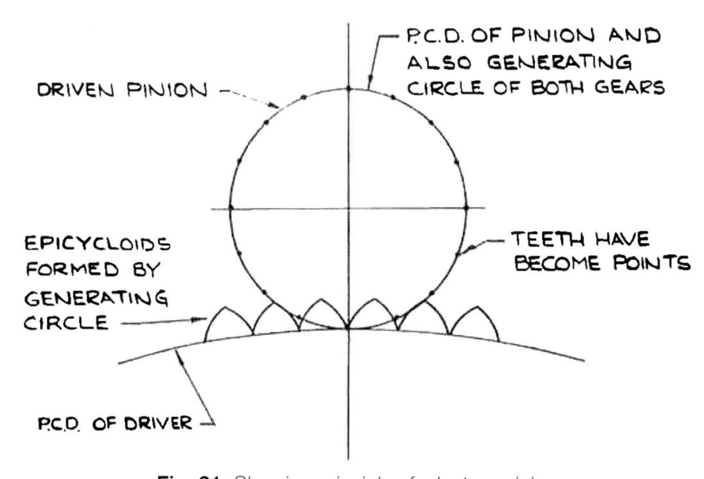

Fig. 21. *Showing principle of a lantern pinion*

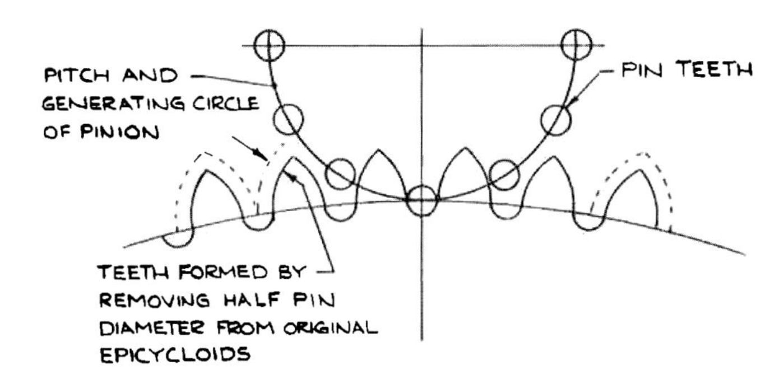

Fig. 22. *Showing the practical form of the lantern pins and teeth are still of correct cycloidal form*

a space will be made available for the pin; however, in order to complete the clearance a semicircular space must be provided below the pitch line of the tooth. Fig. 22 shows the final shape. The pinion cannot exist as loose pins in space and so a pinion is usually a series of pins mounted between two sides

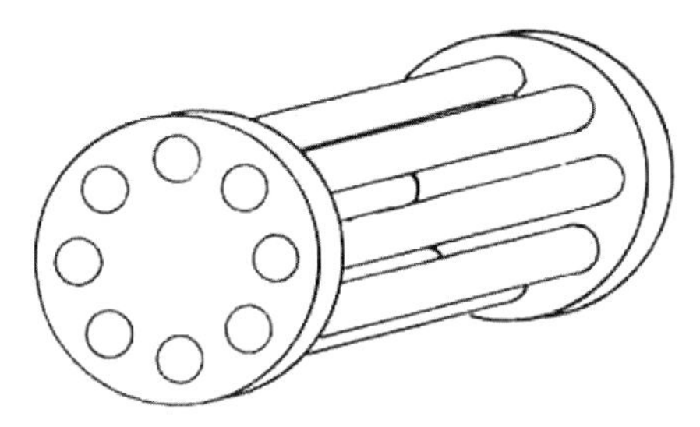

Fig. 23. *Showing the form of a lantern pinion*

or end cheeks. This type of pinion is called a lantern pinion simply because it resembles an old type of lantern—see Fig. 23.

In all gears of this type, if the wheel A, Fig. 20, is the driver then the engagement of the gear teeth will take place after the line of centers has been passed and so is in the low friction area of recess. This results in the minimum of power being lost during transmission. The wheel with the projecting teeth must always drive the pin wheel or lantern pinion. Should the lantern pinion be made the driver then the contact between the teeth would be in the high friction arc of approach and the whole object of the design would be lost. It is not usual in general engineering for the larger wheel to drive the smaller pinion and so the lantern wheel is mainly confined to clock work where this arrangement is the norm.

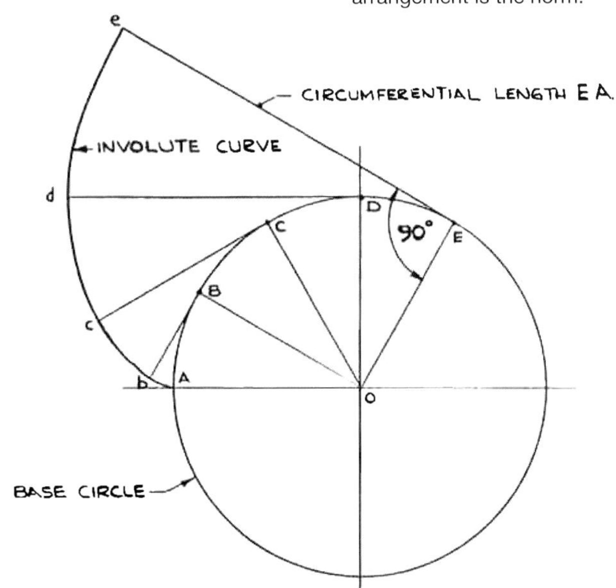

Fig. 24. *The involute curve and its construction*

It can be seen that cycloidal teeth can vary considerably in shape and yet still fulfill the needs of smooth action and constant velocity between the mating gears. So long as the generating circles are the same diameter for any two gears running together, then whatever the resulting shape of the teeth may be, the two gears will run together satisfactorily.

INVOLUTE TEETH

Gears based upon the cycloidal curve do maintain constant velocity between the two PCDs but this is only true so long as the two PCDs are touching each other. If, for any reason, the gear centers are opened out and the PCDs lose contact then the constant velocity condition will not be met. This means that if the wheel centers are not accurately set in the first place or, as is more likely, the wheel bearing wears, thus allowing the gears to spread, the constant velocity condition will be lost. The gear ratio between the gears cannot alter as this is determined by the number of teeth on the respective gears but the gears will not revolve evenly; speeding up and slowing down will take place as the teeth come into and out of the engagement, thus introducing vibration and noise in the mechanism. If the gear teeth are shaped on the involute curve then the constant velocity condition will not be affected by a small amount of spreading of the gear wheel centers, which means that any normal wear in the gear wheel bearings will not adversely affect the proper action of the gear teeth.

The definition of the involute curve as applied to gear teeth is a geometrical curve traced by a point in a flexible, inextendable cord being unwound from a circular disc, the circumference of which is called the base circle, the disc being concentric with the pitch circle of the gear. That is the technical definition of an involute but it is more easily understood by thinking of a disc or drum with a string around it and as the string is unwound from the drum, so long as the string is kept taut, the path taken by the end of the string will describe an involute curve; Fig. 24 shows an involute curve and its geometrical construction. Any number of radial lines, OA, OB, OC, etc., may be drawn and the more lines that are used the more points on the curve will be produced, thus making it easier to draw the involute correctly. The lines Bb, Cc, Dd, are always tangents or, in other words, the angles OEe, ODd, OCc, etc., are always right-angles and the length of the lines Bb, Cc, etc., is the same length as the circumference between the relevant tangent point and the start of the "unwinding" point A.

Unlike the cycloidal curve, which has a definite start and finish and can be repeated over and over again, the involute is one curve only and goes on indefinitely as the unwinding action continues. Although the curve is endless the only part that is considered in the formation of gear teeth shape is the initial part of the curve where it leaves the base circle. Fig. 25 shows a typical tooth shape for an involute gear. It can be seen from this that the contact surfaces of the face and flank are not two curves intersecting on the pitch line but one continuous curve that starts on the base circle. The base circle is not the pitch circle, it is the disc or drum from which the "string" unwinds. The base circle is always smaller in diameter than the PCD but must be concentric to it. The tooth extends below the base circle by a small amount which provides a clearance between the bottom of one tooth and the top of another and also allows a space for a small blending radius to add strength to the root of the tooth.

Any pair or train of involute wheels that are to work together must have the same ratio between the radius of the base circle and the radius of the pitch circle. Reference to Fig. 26 will show that the ratio between the base circle and the PCD is not usually quoted as a ratio but as an angle. AC is the center line of the tooth passing through the center of the gear and where this line intersects the pitch

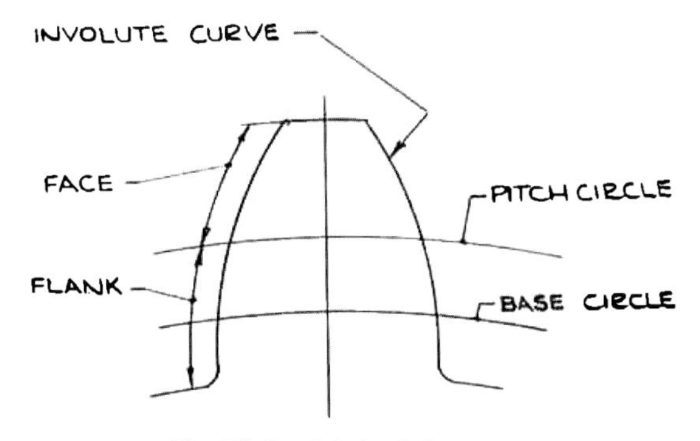

Fig. 25. *Involute tooth form*

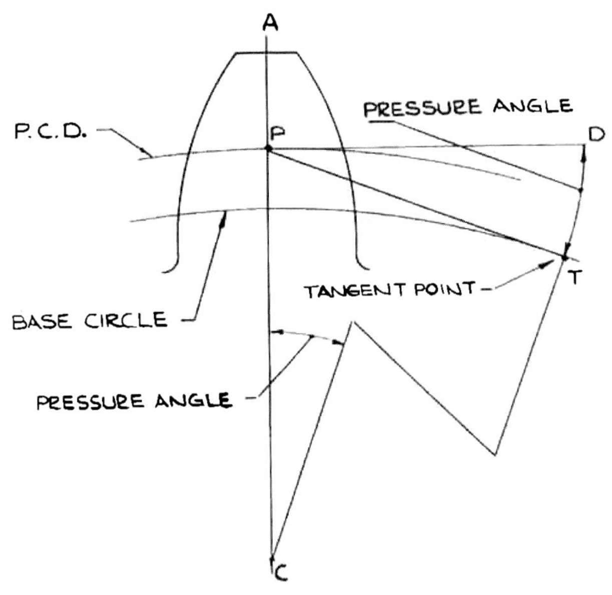

Fig. 26. *This shows how the ratio between the base circle and the PCD is represented as an angle*

circle establishes the position of point P. A line is now drawn from point P tangential to the base circle, the tangent point being point T with the angle TCP being the pressure angle of the gear. It is more usual, however, to show angle TPD on gear tooth drawings as this has the same value as angle TCP but eliminates the necessity of drawing the gear center. In the case of the cycloidal teeth it was seen that the line of contact between the gear teeth was a combination of two curves; in the case of the involute form the line of contact is a straight line and is shown in Fig. 27 as the line AB. All the contact takes place outside the base circles.

Altering the pressure angle of the tooth not only alters the shape of the teeth but also alters the length of the contact path: the greater the pressure angle the longer will be the line of contact. Fortunately in modern involute gearing there is not a great proliferation of pressure angles, the most common angle in use being 20°. This angle has been found to give good tooth shape and

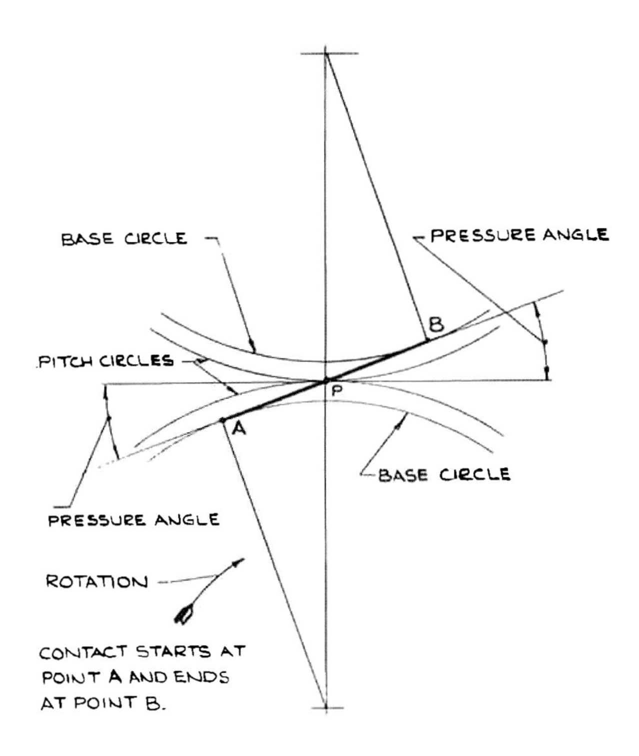

Fig. 27. *The line of contact between two involute teeth*

also provide a strong tooth root, producing a gear that is capable of long life and good performance. The only other pressure angle that has been in general use is 14½°, however, this angle seems to have fallen out of favor with gear designers although some lathe manufacturers still use the 14½° pressure angle for the change-wheels.

The larger pressure angles make the teeth more "stubby," while the smaller angles produce teeth that appear longer. Only gears of the same pressure angle should be used in the same train. Gears with a 20° pressure angle should not be used to mesh with gears based on the 14½° angle; although they would rotate their action, performance and life would be seriously impaired. Any involute gears of the same pressure angle and pitch will run together correctly although the shape of the teeth on one gear may not be similar to the shape of the teeth on the mating gear. This is due to the involute curve varying with the diameter of its base circle. The tooth shape will have a pronounced curve on a pinion possessing relatively few teeth, while on the larger wheels the sides of the teeth may be almost straight. In fact, the rack form of the involute is a perfectly straight line at the pressure angle.

CHAPTER 3

GEAR TOOTH SIZES

In the previous chapters the shape of the gear teeth has been discussed but not the size. Obviously a gear of, say, 3" diameter could have 20 teeth upon it or it could have 200 teeth. Both sets of teeth could be of perfect involute form but they could not be run together as their physical sizes would be considerably different. Some means or standard of determining tooth size is therefore necessary. The basic datum for all gear calculations has up to now been the pitch circle diameter and it remains so for determining or defining the tooth size. There are three methods in general use for specifying the size of gear teeth, these are the circular pitch, the diametrical pitch and the module. If any one of these is known then the others, if required, can be determined by simple calculation.

CIRCULAR PITCH

The circular pitch of a gear is the distance from a point on one tooth to a corresponding point on the next tooth measured around the pitch circle diameter. Fig. 28 illustrates this. The circular pitch is usually quoted as some round figure, such as ¼", ½", ¾", etc., however, using a round number for the circular pitch will invariably result in an awkward number for the pitch circle diameter. The reason for this is that the circumference of the PCD equals the number of teeth in the gear multiplied by the circular pitch. The number of teeth must, of course, be a whole number and when this is multiplied by the round figure of the circular pitch it will naturally result in a round figure for the circumference. However, in order to obtain the PCD the circumference must be divided by pi, and any round figure divided by 3.1416 will in itself be awkward. It can be argued that in the workshop one size is

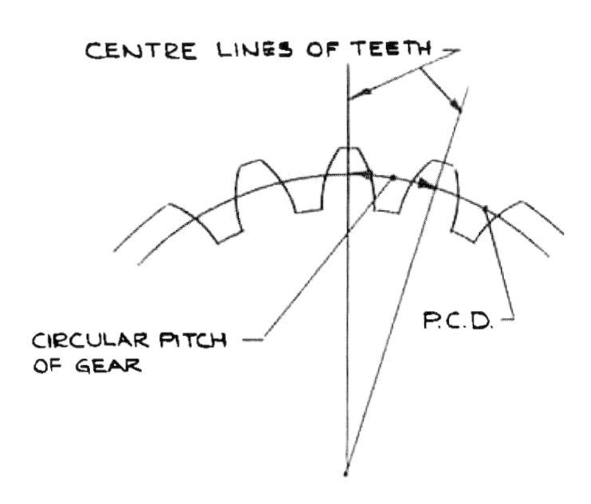

Fig. 28. *Showing how circular pitch is measured*

just as easy to measure as any other and that measuring by a vernier caliper gauge to 3.017 is just as easy as measuring 3.000. This may be so but it certainly helps the designer, particularly when setting out a train of gears, to be able to use round figures for the wheel centers. The problem is largely academic because the relatively small gears that the amateur will encounter will not be based on the circular pitch notation. The exception to this could be in things like downfeed pinions for drilling and milling machines.

DIAMETRAL PITCH

This is certainly by far the most common and the most useful method of notation for small gears and the definition of a small gear in this case is any gear that the model engineer or backyard amateur is likely to handle. The diametral pitch is simply the number of teeth a wheel has per inch of pitch diameter. For example, if a gear has a PCD of 2" diameter and has 40 teeth then it is said to be 20DP. If the number were to be 40 on a PCD of 1" diameter then the DP would be 40. DPs are usually whole numbers and, more often than not, even numbers. It is not usual to encounter an odd DP number after 10DP has been reached and gears below 10DP will in the main be larger than the amateur will want to cut. This arrangement makes the setting out of gear train centers easy. For example, should it be decided to produce two gears to give a ratio of 3:1 and 20DP was chosen for the tooth size, two gears—one with 20 teeth and one with 60 teeth—would appear to be satisfactory. The PCD of the 20 tooth would be 1" while the PCD of the 60 tooth would be 3". This would mean that the gear center would be 2", or half of the addition of the two PCDs.

MODULE

Older textbooks quote the module as being the reciprocal of the DP but more recently it has become the "metric" way of quoting the

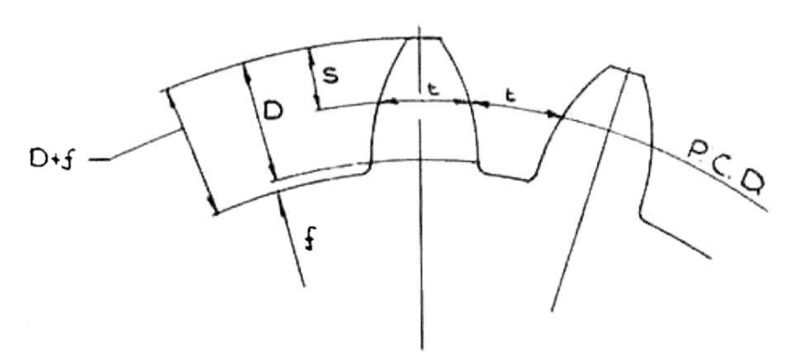

Fig. 29. *Tooth proportions*

size of the teeth. The module can be said to be the pitch diameter in millimeters divided by the number of teeth, or to put it the other way around, the PCD in millimeters is the module number multiplied by the number of teeth in the gear. As there are 25.4 millimeters to one inch then a number 1 module is equal to 25.4DP, a number 2 module would be 12.7DP, while a .5 module would be 50.4DP.

TOOTH PROPORTIONS

As yet nothing has been said about the proportions of the teeth, that is the amount the teeth project above or below the pitch circle and the width of the teeth at the pitch circle. These proportions are shown in Fig. 29 but this is really for academic interest only as when cutting teeth the only control the operator has over tooth proportion is the depth to which he cuts the teeth. Provided the teeth are cut to the correct depth the shape of the cutter will determine the proportion of the teeth.

The whole depth, or the cutting depth of the teeth is shown by the symbol D+f. This is the international symbol and for the benefit of the operator is usually quoted on all commercial cutters. Fig. 29 also shows the thickness of the tooth at the pitch line to be the same as the tooth space at the pitch line. While this is theoretically correct, in practice the tooth space is usually made a little wider than the tooth thickness. Some textbooks quote the tooth thickness as being .48 of the circular pitch, thus making the tooth space .52 of the circular pitch.

CHAPTER 4

RACK AND PINION GEARS

Rack and pinion gears are not usually thought of as being two ordinary rotating spur gears but, in actual fact, that is basically what they are. A rack is only a small part of a very large spur gear wheel. If the diameter of a gear were to be increased in size to infinity then its pitch circle diameter would become a straight line. A rack is only a small section cut out of this gear of infinite diameter.

Rack and pinions only differ in operation from a pair of spur gears in the specific duties they perform. Two spur gears, because they are circular, could keep rotating for an indefinite number of revolutions. In the case of a rack and pinion the rack has a predetermined travel and once the pinion reaches the end of the rack, motion has to cease and further movement has to be in the opposite direction so that at the end of each cycle of operation both rack and pinion return to their original starting position. One consequence of this is that the same tooth on the pinion always engages with the same tooth on the rack. This condition only applies to rotating gears when both driven and driver gears have the same number of teeth. A rack and pinion is a means of converting linear motion from rotary motion or, when the rack is the driver, rotary motion from linear motion.

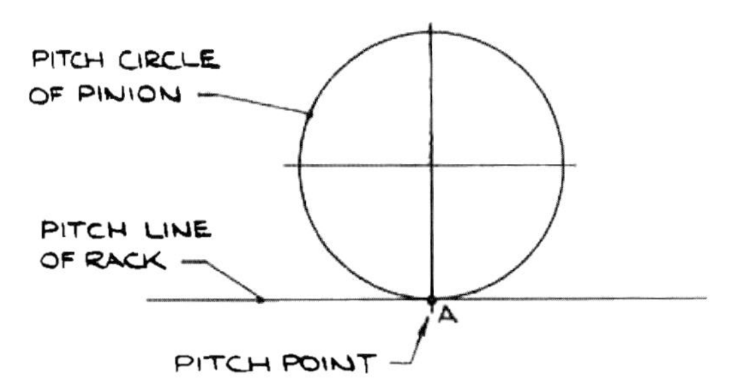

Fig. 30. *Showing how the principle of the pitch circle is applied to a rack and pinion*

The two gears are set out and planned in accordance with the same principle already outlined for designing a pair of spur gears. In Fig. 30 the pinion P is represented by its pitch circle diameter and the rack by a straight line, this being a portion of a PCD of infinite diameter. The two pitch lines contact each other at point A. In practice just consider the PCD as a disc rolling along the straight pitch line. The length of the rack required will naturally depend upon the diameter of the PCD of the pinion. Every revolution of the pinion will require a rack with an effective length equal to the circumferential length of the pinion. A pinion with a PCD of, say, 2" will require about 6½" length of rack for each revolution the pinion has to make to fulfill its design requirements.

All the aspects of tooth shape that were discussed with regard to spur gears must also be applied to rack and pinion teeth if the constant velocity of the mating pitch line is maintained. The gear tooth form can be either involute or cycloidal, either standard giving satisfaction where the teeth are correctly designed and made.

THE INVOLUTE FORM OF RACK AND PINION

If the involute form has been employed for a rack and pinion then both the rack or the pinion can be used indiscriminately as the driver or driven component and either configuration will perform equally well. The involute shape of the teeth on the pinion will be exactly the same as that used on a spur gear, in fact any spur gear of the same pitch and pressure angle as the rack will mesh correctly with it, which is not surprising because it has already been established that the rack is only a part of a large spur gear.

The rack teeth will differ slightly from the pinion teeth owing to the PCD of the rack being of infinite size. The shape of the resulting involute will be a straight line. The rack teeth will therefore be straight-sided throughout their entire working length. The teeth will not be perpendicular to the pitch line but inclined at the pressure angle and so the rack will have a form as shown in Fig. 31.

The pitch of the rack teeth is measured from any point on one tooth to a similar point on the next tooth—in Fig. 31 the center line of the

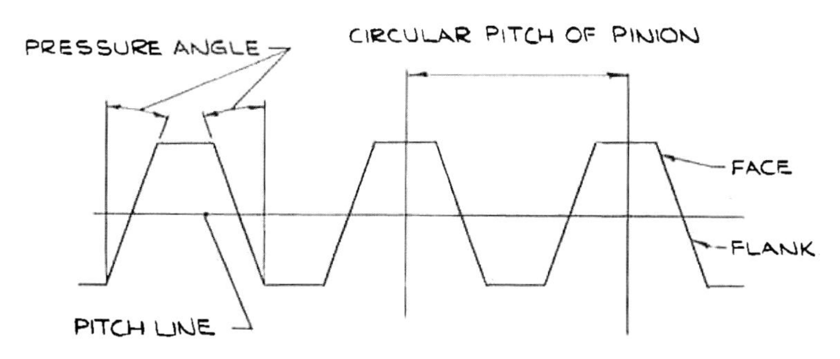

Fig. 31. *Involute form of rack*

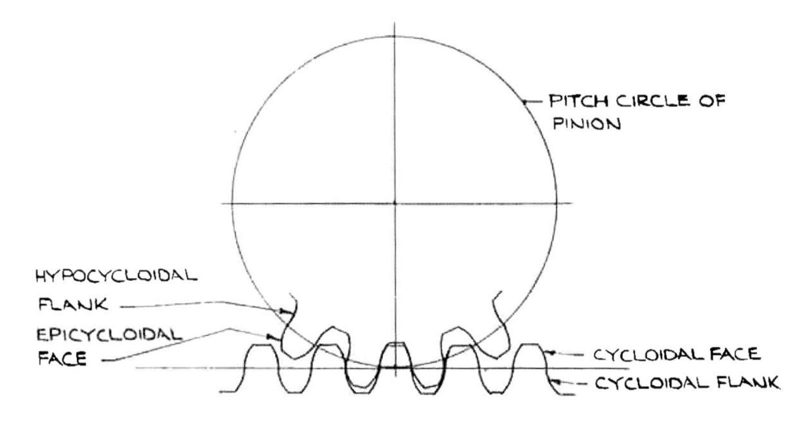

Fig. 32. *Cycloidal form rack and pinion*

tooth has been chosen. Usually the circular pitch method of denoting the size of the tooth is used as this gives a nominal round figure for the pitch of the rack and although this results in the pinion having an awkward PCD, that is generally preferred to a non-standard rack pitch. If a rack is used to provide a feed for a machine tool such as the downfeed of a vertical milling machine, it becomes easier to calibrate a dial for a pitch of .200 than it would be if a 16DP standard had been used, which would result in a rack pitch of .1963.

RACK AND PINIONS WITH CYCLOIDAL TEETH

The cycloidal tooth form can be applied to a rack and pinion in just the same way as was used in spur gearing. Should the design requirements demand a rack and pinion where either could be called upon to be the driver, then the system of epicycloidal faces and hypocycloidal flanks should be used for the pinion teeth. The rack teeth will be a shape different from any previously described.

The PCD of the rack is a straight line and the generating circle rolling upon it will produce a true cycloid as shown in Figs. 13 and 14. The tooth shape above the pitch line, or the face, will therefore be a cycloid and not an epicycloid as they have been on other occasions. The curve produced by rolling the same generating circle underneath the pitch line will again be a cycloid, in fact it will be a mirror image of the first one, so the tooth below the pitch line will also be a cycloid and not a hypocycloid. The final tooth shape will be a combination of two similar but "opposite hand" cycloids blending together at the pitch line. Fig. 32 shows the shape of both pinion and rack.

PIN TEETH APPLIED TO RACK AND PINIONS

The teeth of the rack or pinion may be of the circular or pin form, employing the same principle used in the lantern pinion and wheels shown in Chapter 2 and illustrated in Fig. 22. In this case the duties of the two components cannot be interchanged. The

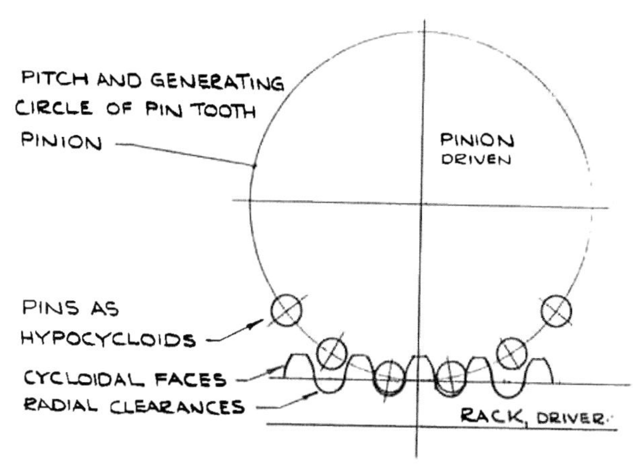

Fig. 33. *Rack with lantern pinion: the rack is the driver*

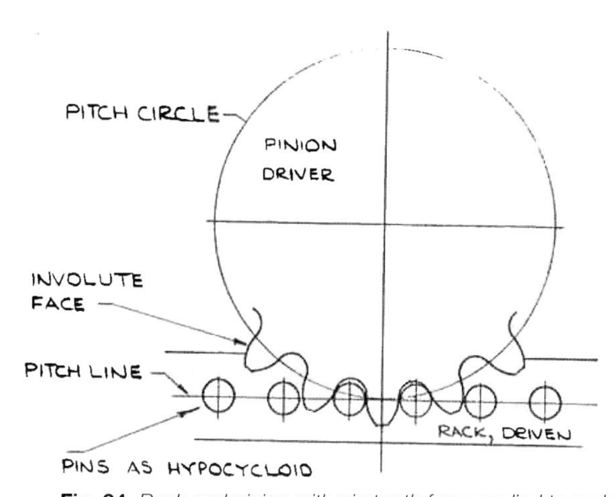

Fig. 34. *Rack and pinion with pin tooth form applied to rack*

pin teeth must always be on the driven component and if the pin wheel is used as a pinion then the rack must be the driver. Should the pin teeth be on the rack then the pinion must be arranged to drive the rack.

The arrangement whereby the pinion becomes a pin wheel is shown in Fig. 33. The pins represent the hypocycloid produced by using a generating circle whose diameter is the same as the pitch circle diameter of the pinion. The teeth on the rack will have a true cycloidal face above the pitch line, the cycloid being formed by rolling the PCD of the pinion along the straight pitch line of the rack. The radial spaces below the pitch line are only to allow a clearance for the passage of the pins and take no part in tooth engagement.

All tooth engagement takes place after they have passed the center line and are in the low friction area of contact but this applies only so long as the rack is the driver. The actual width of the rack will have to be less than the distance between the two end plates of the lantern pinion. This is necessary in order to provide a clearance for the end plates whose diameter for practical reasons must be greater than the PCD of the pinion and so will extend well below the surface of the rack.

In cases where it is desirable to have the pinion as the driver, the pin teeth must be applied to the rack and this leads to a curious situation. The pin teeth represent a hypocycloid and, as before, the generating circle must be equal in diameter to the PCD, but the PCD of the rack is of infinite diameter, which is a straight line! This does not affect the validity of the pins being hypocycloids—the theory of the pins holds good for any PCD and generating circle that are the same size irrespective of what, or how great, that size may be.

What is affected is the teeth on the pinion which will have only faces and no flanks since, of course, the faces are formed by a generating circle that is a straight line. This straight line, rotating or rocking on the PCD of the pinion, will in fact behave exactly as the "flexible inextendable cord" used in the definition of the involute curve, and the resultant shape of the epicycloid will be an involute curve. We have therefore the strange situation of the faces on the pinion teeth being of involute form yet they are engaging quite correctly with the hypocycloidal pins on the rack. Fig. 34 illustrates this situation. The end plates supporting the pin teeth of the rack will naturally encroach over the pitch circle of the pinion. The face width of the pinion must therefore be less than the distance between the end plates of the rack to allow for a working clearance.

CHAPTER 5
BEVEL WHEELS

Should the occasion arise where it becomes necessary to transmit power from one shaft to another and where the two shafts are not parallel to each other then the use of bevel wheels will provide a satisfactory solution. From the amateur constructional point of view, the teeth on bevel wheels—unlike spur gears—are not easy to cut. The shape of the teeth is such that without specialized equipment, not usually found in the normal home workshop, the correctly formed bevel tooth cannot be produced. However, there are ways whereby a compromise on tooth form can be achieved and bevel gears produced that may satisfy the constructor's requirements, although the actual tooth form will be an approximation rather than the correct theoretical shape.

The problems involved will be better understood by considering the action of a pair of bevel wheels in a similar way to that used with spur gearing, that is as a friction drive without the teeth. In this case the rotating disc will have to be replaced by rotating cones. Where the two shafts are at right-angles to one another and rotate at the same speed the resultant gears are sometimes referred to as miter wheels: a cone type representation of this arrangement is shown in Fig. 35. If we consider the largest diameters of the two cones CC and DD we find they will be identical in size owing to the two cones being identical and it follows that one revolution of diameter CC will result in one revolution of diameter DD and so both diameters will rotate at the same speed.

Any other two diameters of the cones, such as EE and FF, or GG and HH, will be in the same proportion to each other as the base diameter and so their speeds of rotation will be exactly the same as the base diameters. No matter where a section of cone A is taken, so long as it is parallel to the base and at right-angles to the center line, its diameter will be the same as the respective section of cone B. From this it can be seen that the entire surfaces of the two cones will roll together without slip. Any two touching diameters, such as EE and FF, could be chosen for the pitch circles of the gear without affecting the relative speeds of the two shafts.

In the case of spur gears it was seen that if the two shafts joined by a pair of gears had to rotate at different speeds then the ratio of the respective gear pitch circle diameter must be similar to the speed ratio required. Similar conditions apply to bevel wheels; if the two shafts joined by the bevels are to rotate at different speeds, then the diameters of the cone bases must be proportional to the

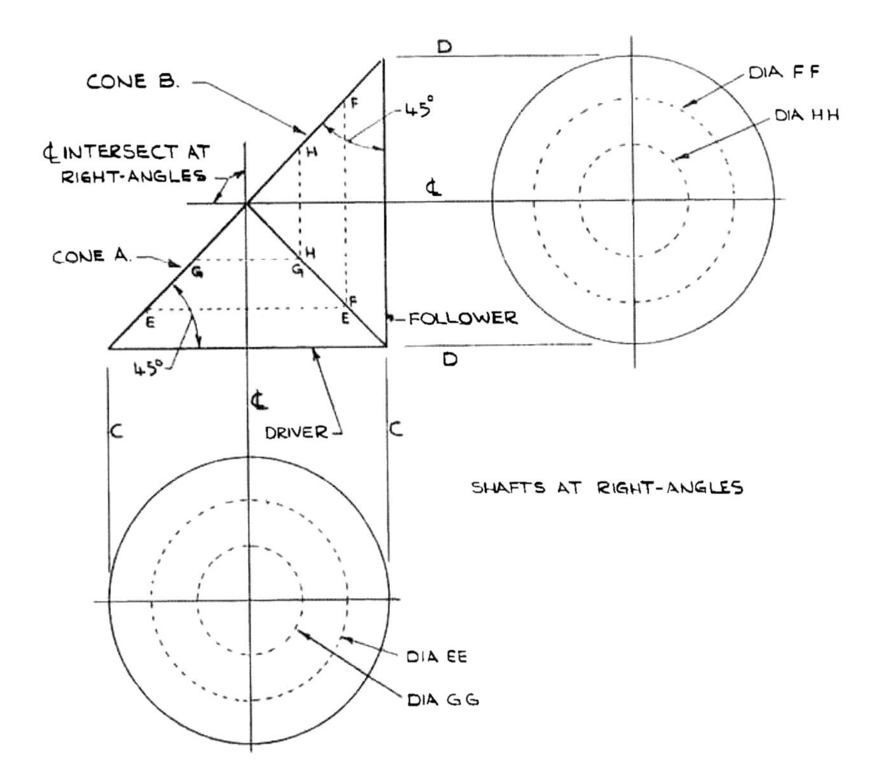

Fig. 35. *A cone representation of a pair of miter gears*

speeds required. This condition is shown in Fig. 36, where the shaft of cone A is to make two revolutions while the shaft of cone B only makes one revolution.

To achieve this the base of cone B has a diameter equal to that of twice the diameter of cone A. Although the two cones are of different diameters and face angles, sections taken at any point up the cone, such as EE and FF, will still give touching circles that have precisely the same velocity ratio as the base circles of the cones. So once again the condition is met whereby the surfaces of the two cones will roll together perfectly without any slip taking place. It will be seen from both Figs. 35 and 36 that a line joining the two cone center-lines is one continuous straight line. This will always be so with shafts at right-angles irrespective of the gear ratios chosen. All the above comments are only true where the two axes of the cones and shafts intersect, but this is the condition that applies to the majority of bevel gears encountered in general engineering practice.

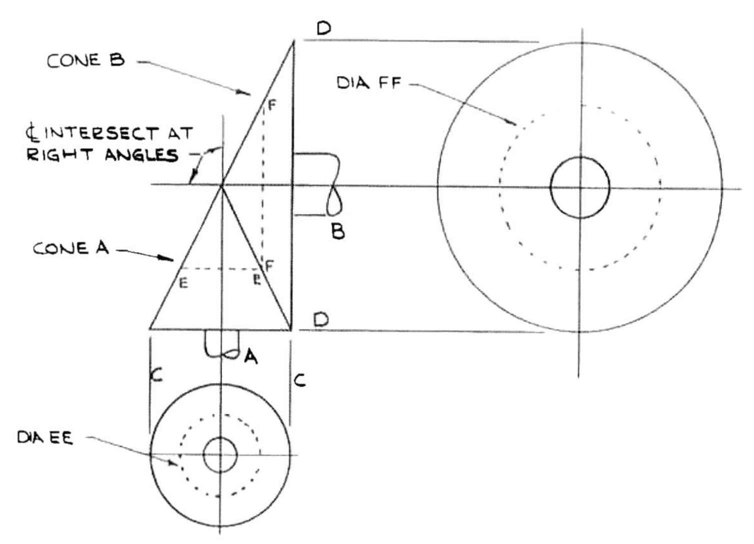

Fig. 36. *Bevel wheels rotating at different speeds but with shafts at right-angles*

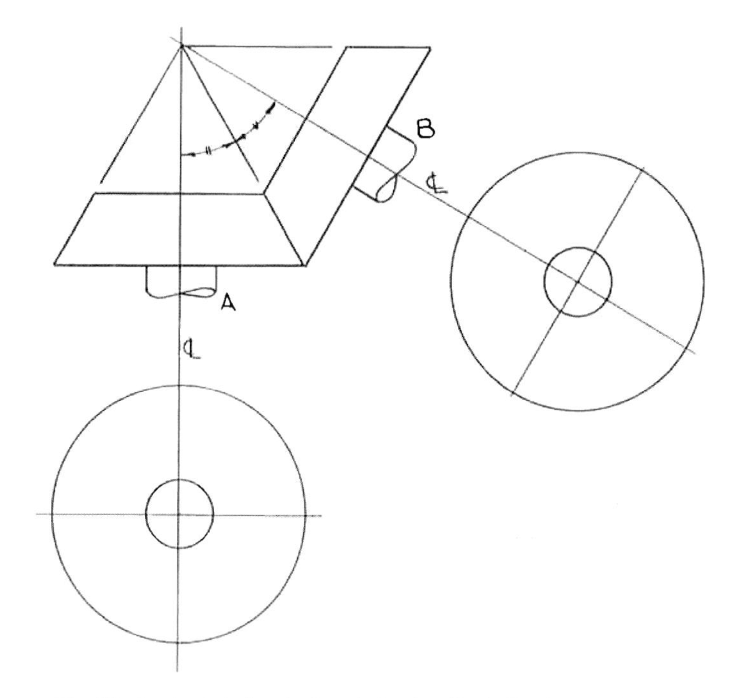

Fig. 37. *A pair of similar size bevel wheels with shafts at an acute angle*

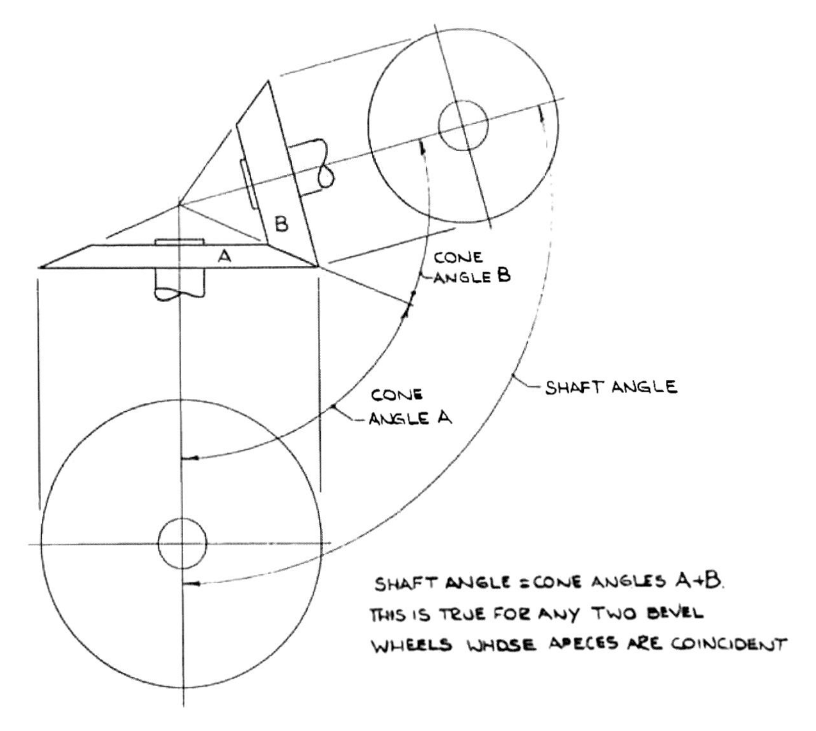

Fig. 38. *Showing a pair of dissimilar size bevel wheels whose shafts are at an obtuse angle*

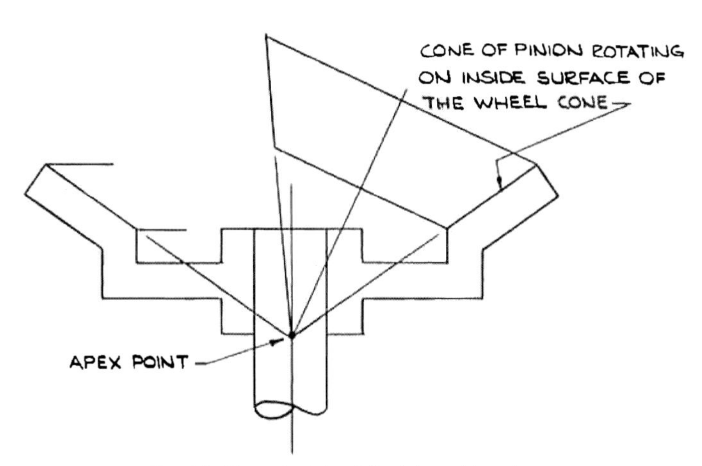

Fig. 39. *The principle of the internal bevel gear*

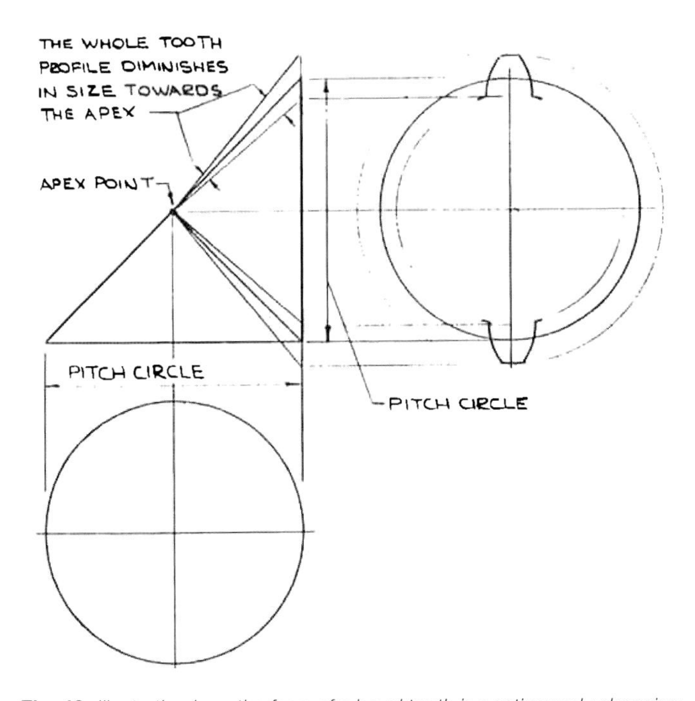

Fig. 40. *Illustrating how the form of a bevel tooth is continuously changing*

Shafts not at right-angles can also be joined together by bevel gears and the principles outlined previously still apply. Fig. 37 depicts two shafts at an angle of less than 90° joined together by a pair of bevel cones. Should the required angle be greater than 90° then the cone configuration will be as shown in Fig. 38. If the two shafts are to rotate at the same speed then the cones will be of similar size, as shown in Fig. 37. If the shafts are to rotate at different speeds then one cone will be larger than the other, see Fig. 38. The base circles of the cones are arranged in the same proportion to the respective speed ratio required, just the same way as was described for the right-angle shaft.

In theory it is possible to have internal bevel gears, that is, where the outside of the cone surface of the pinion will roll in contact with the internal surface of a larger wheel. This arrangement, even in the modern technological age where sophisticated gear cutting machinery is available, is avoided if at all possible as the problems involved in cutting the teeth on the internal cone are considerable. Fig. 39 illustrates this arrangement.

BEVEL GEAR TEETH

All the conditions that were discussed about maintaining constant velocity of the two mating discs when applying teeth to them to form spur gears also applies to bevel gears. When the teeth are added the resultant bevel gears must continue to roll together as though they were still smooth conical surfaces. This introduces a problem as both teeth and spaces must be conical and follow the shape of the pitch surface of the cones. All the surfaces of the teeth, if carried on towards the cone apex, must meet there and so form a point. The size of the tooth in all its aspects is not constant: the whole of the tooth form is continuously changing and diminishes as the cone center is approached. This is shown in Fig. 40. Every aspect of the surface of the teeth, faces, flank, as well as top and bottom—must be conical.

Bevel gear teeth are almost always of involute form and so the actual involute shape will also alter as the tooth profile moves along the conical surface. This means that the normal type of form cutter will not give the correct tooth profile as the cutter would only be of the correct shape at one point along the length of the tooth. Should the teeth of uniform section be used, similar to spur gears, or even if the incorrect cone angle were used so that the teeth did not terminate at a point at the intersection line of the two pitch cones, then the gears would not work together correctly. It may not even be possible to assemble the gears at their correct positions as the teeth would interfere with each other rather than dropping into correct mesh. Should power be applied to rotate this type of gear then the teeth would rapidly break up under the high stress imposed upon them.

In practice the teeth are not carried on up to the apex point because as the center is approached the teeth become smaller until they finally disappear altogether. Only a portion of the length of the tooth is of practical value for transmitting power and this is naturally at the large end of the cone. There is no definite fixed standard for tooth length but it is usually about one-third of the total length of the cone surface, Fig. 41.

In order to give the correct rolling contact and maintain constant velocity between the two cones, the teeth on bevel wheels are formed using the same principle described for spur gear teeth. There is, however, one considerable difference; with spur gears the teeth are formed on the actual pitch circle diameter, but this is not so with bevel gears. Referring to Fig. 40 it can be seen that the teeth are not in a plane parallel to the base of the pitch cone but are, in fact, perpendicular to the conical pitch surface. It would therefore be quite wrong to develop the teeth on the circumference of the pitch circle of the cone as this circle is perpendicular to the cone's axis. The involute curve has to be developed from a circle perpendicular to the conical surface of the cone. Again referring to Fig. 40, the shape of the teeth shown in the end view is not the true shape of the teeth but the shape that is seen by looking square with the cone's axis rather than square with the cone surface.

This may be better understood by referring to Fig. 41 where the teeth have been added to the cone. The pitch circle of the cone is shown by the line BB and the apex is point A. The length of the pitch cone is the line BA, to which teeth have been added and are shown shaded on the drawing, the face of the tooth being above the pitch line while the flank is below. Both the top and the bottom of the tooth, if projected, would meet at the apex point. The teeth are formed looking in the direction of arrow F which is looking up the surface of the pitch cone. If a line is drawn at right-angles to the pitch cone and projected

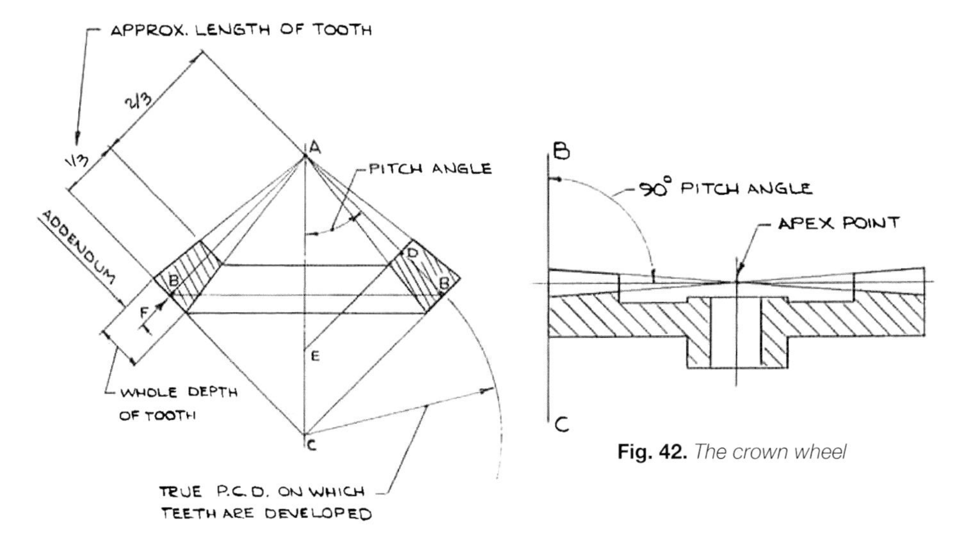

Fig. 41. *Showing the true pitch circle diameter of the teeth*

Fig. 42. *The crown wheel*

back until it meets the cone center line, point C will be established. The length of the line BC is the radius of the true pitch circle diameter on which teeth are developed. Similar radii, such as ED, can be established for any point along the length of the tooth.

CROWN WHEEL AND PINION

The term crown wheel and pinion is often used to describe a pair of bevel wheels where the pinion is small compared to the wheel; a gear ratio of around 4:1 would result in the wheel appearing large when placed at the side of its mating pinion and may merit the term crown wheel and pinion. This is, however, not strictly correct. A crown wheel is a special type of bevel wheel where the pitch angle has been increased to 90°. By following the method already described to determine the diameter of the true pitch circle it will become apparent that the line BC, Figs. 41 and 42, will never intersect with the center line of the cone, which means that the true pitch circle is of infinite diameter.

The crown wheel is, therefore, the bevel wheel's equivalent of the rack. In the case of the rack, the straight line pitch circle resulted in the involute becoming a straight line and the same applies to the crown wheel, the teeth have straight sided faces and flanks but

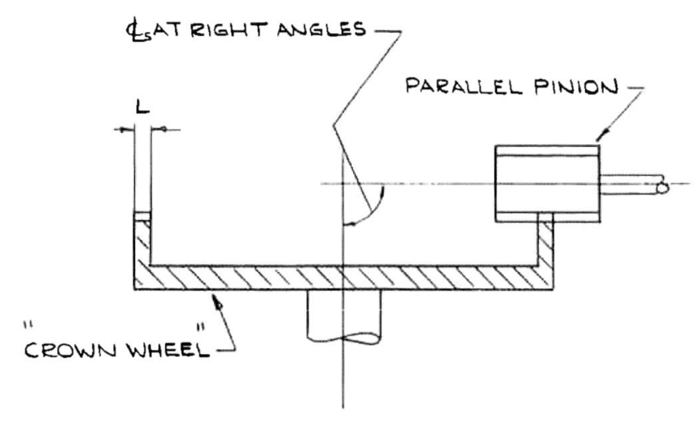

Fig. 43. *A "clockwork" crown wheel and pinion*

all features of the tooth must still meet at the center line of the wheel even though there is no cone. Fig. 42 shows a crown wheel and how the teeth still taper towards the center. The crown wheel, or to be more correct, the theory behind it, is used in industry as the basis of the generating system used in bevel gear production.

Occasionally, and particularly in clockwork mechanisms or other low power transmissions, the "crown wheel and pinion" configuration shown in Fig. 43 may be used. The teeth of both wheel and pinion are not conical but parallel over their entire length. This is theoretically quite incorrect and only works in practice where the length of the tooth on the so-called crown wheel is at a minimum and the ratio between the two gears is large, thus resulting in a large wheel and very small pinion.

To understand the basic error in the arrangement refer to Fig. 44. Here the teeth of both gears have been removed, the wheel being represented by a flat disc and the pinion by a cylinder. The pinion is of constant diameter and as its end is in contact with the wheel at diameter A it must move at the same speed as the other end in contact with diameter B on the wheel. The pinion will try to drive the wheel at diameter A at a speed greater than it can at diameter B. This condition is true at any diameter of the wheel between A and B that is in contact with the pinion. Since all diameters of the wheel must rotate at the same speed, a slipping action must take place between the two surfaces in contact. As this slip could not occur when the teeth are added to the two components, the whole arrangement is impractical and can only be made to work by reducing the tooth length (L in Fig. 43) to little more than a raised ridge. The wheel is only then able to transmit small amounts of power otherwise the teeth would rapidly wear and totally fail.

LAYING OUT A PAIR OF BEVEL WHEELS

It is certainly true that the setting or laying out of a pair of bevel wheels is not so simple a process as laying out a pair of spur gears, but nevertheless it is not particularly difficult if the correct procedure is followed. In the case of spur gears which connect parallel shafts, the size of the wheels is limited by the distance between the two shaft centers. When the shafts are angled this limiting factor does not occur so the bevel wheels may, in theory, be any size, thus giving the designer considerable latitude in choosing both the size of the wheels and also the number of teeth to be used. A large pair can naturally employ bigger and stronger teeth than a pair chosen farther up the surface of the generating cones. Probably the best way to illustrate the method of determining the actual shape and size for a pair of bevel wheels is to work through an example.

Supposing therefore that a pair of shafts at right-angles are to be connected by a pair of gears. The shafts are to have a speed ratio of 2:1 and 20DP teeth are to be employed. There is a wide choice available in determining the size of the gears, as any two gears that have numbers of teeth in the ratio of 2:1 will meet the speed requirement of the two shafts. For this example we will choose a 60-tooth gear and a 30-tooth gear. The layout is shown in Fig. 45.

First draw in the two shaft centers at right-angles, these are the lines AA for the gear and BB for the pinion. Next, draw lines for the cone base or pitch diameter; in the case of the wheel the diameter will be 3"—this is found by dividing the number of teeth, which is 60, by the DP. A similar calculation on the pinion will give 30 divided by 20, which is 1½". These lines are shown as CC and DD for the gear and EE and FF for the pinion.

The pitch lines can now be drawn, these pass through the intersecting points of the cone base lines and also pass through the center point X. The pitch lines are shown as

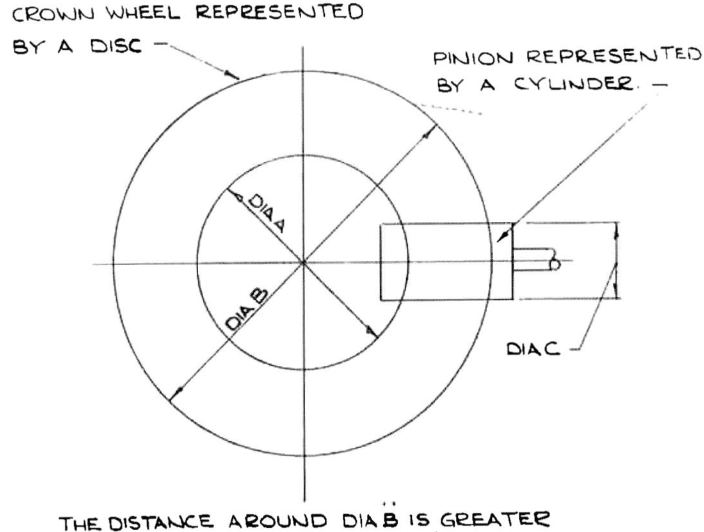

Fig. 44. *Showing why the gear depicted in Fig. 43 is incorrect*

GG and HH. Lines at right-angles to the pitch lines are drawn next, these lines passing through the intersection points of the pitch lines and cone base lines and continuing until they intersect the cone center lines at Z and Y. These lines are sometimes referred to as the back cone lines and represent the back of the teeth.

The teeth can now be drawn in. The height of the addendum is 1 divided by the DP, in this case 1⁄20, or .050, the whole depth of the tooth being twice this amount—although this is not strictly correct as no clearance has been shown but it is nevertheless satisfactory for layout purposes. The lines representing the teeth should, of course, continue until they all meet at the apex point X. The length of the teeth will be influenced by the load they have to carry and is a matter of general designing, as is the rest of the outline of the gears. The outline shown in the illustration represents a typical bevel gear shape.

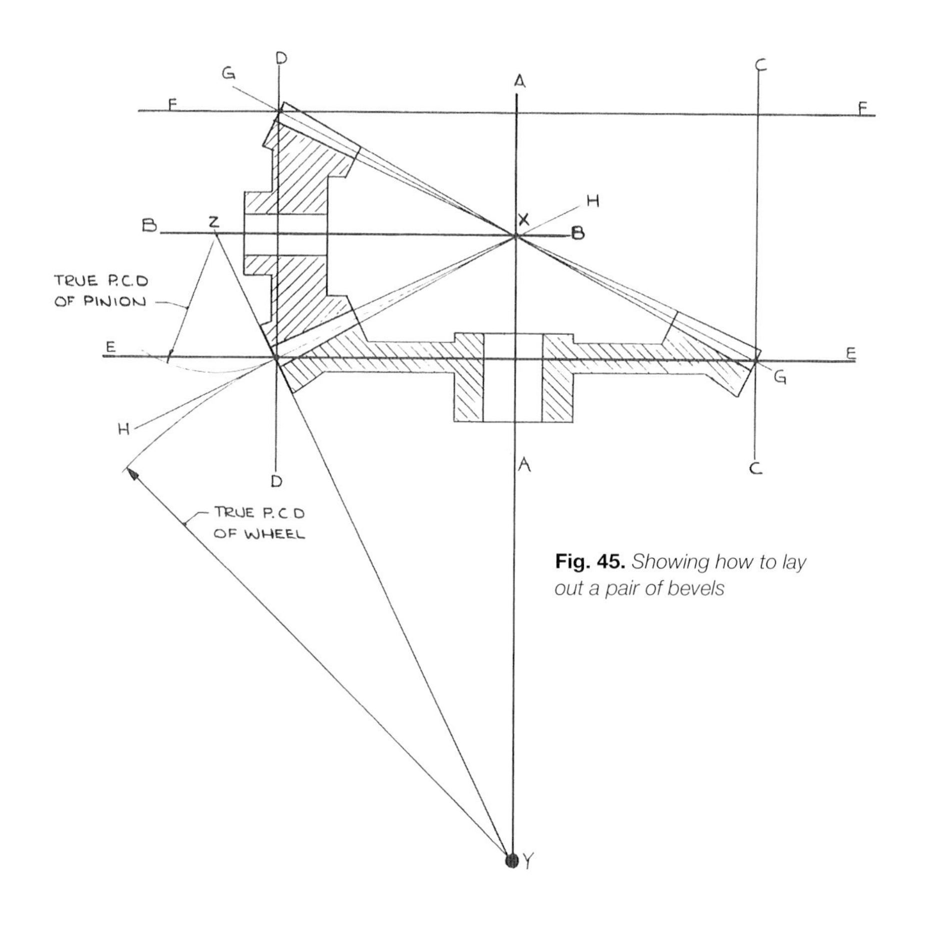

Fig. 45. *Showing how to lay out a pair of bevels*

CHAPTER 6
WORM GEARS

There are occasions when a large gear reduction is required and to obtain this by means of spur gearing would necessitate quite a large number of shafts and gears which together with their attendant bearings may result in a cumbersome and heavy gear box. Although in theory there is no upper limit to the ratio that can be obtained by a spur gear and pinion, in practice the wheel soon becomes too large to be accommodated in the space that can be made available for the complete gear box. It is not usual to have a pinion with much less than 20 teeth and so to obtain even a modest gear ratio of 5:1 would require a wheel having 100 teeth and this could well be too large to be acceptable.

In this case a further shaft would have to be introduced and two more gears employed, one pair to obtain a ratio of, say, 2½:1 and the other pair to provide a 2:1 reduction. The gear train would have to be compounded with two of the gears secured to one shaft—compounding will be explained later when dealing with lathe change wheels. This shows that considerable problems would arise should spur gears be used to provide a high gear ratio of, say, 60:1. Spur gears are not suitable for reductions of that magnitude so some other form of gearing must be found and the solution is to use a system known as worm gearing. With a worm gear a ratio of 60:1, or much larger if required, can be achieved with just two gears, one being a worm and the other a worm wheel. With spur gears either wheel of a pair will look similar to the other, the only difference being in physical size and it is possible for either gear to drive or be driven by the other one. With worm gears the two components are in no way similar in appearance, and although the worm can always be the driver, in most cases the worm cannot be driven by the worm wheel.

A worm and wheel are shown diagrammatically in Fig. 46. The worm, which is the driver, is basically a screw thread engaging into teeth on the periphery of the wheel. The wheel therefore is a gear and resembles a spur wheel in appearance. If the wheel were to be fixed so that it could not rotate, then as the worm rotated it would move along its line of axis in a similar way to a screw being turned into a nut. In fact the teeth of the worm wheel would be acting in a similar way to the threads of a nut. The worm would move either backwards or forwards depending on whether it was being rotated clockwise or counterclockwise.

Should the worm be prevented from axial movement and the wheel made free to rotate on its bearings then as the worm rotates, since it cannot now move axially, the movement would be imparted to the wheel and so the wheel would now rotate. If the direction of rotation of the worm were to be reversed then the direction of the wheel would also be reversed. The rotation of the worm and its shaft would be transmitted to the wheel and so onto the shaft of the wheel, Fig. 46.

In understanding and designing a pair of spur gears it can be seen that they must be thought of as a pair of rotating pitch circles. It is precisely the same with worm gears but with one big difference: the pitch circles or pitch surfaces of spur gears will roll together and theoretically drive each other by means of friction, but this is not so with worm gearing. Fig. 47 depicts a worm and wheel as touching pitch surfaces. If the worm is rotated there will be no turning effect imparted on the wheel at all. The worm will try to move over the surface of the wheel parallel to the axis of the wheel but since this movement will be prevented by the bearings of the worm, the two touching surfaces will rub or grind together and give rise to excessive friction.

Conversely, if the wheel disc is rotated all the energy will be expended in the line of the worm shaft and again no turning effect will be

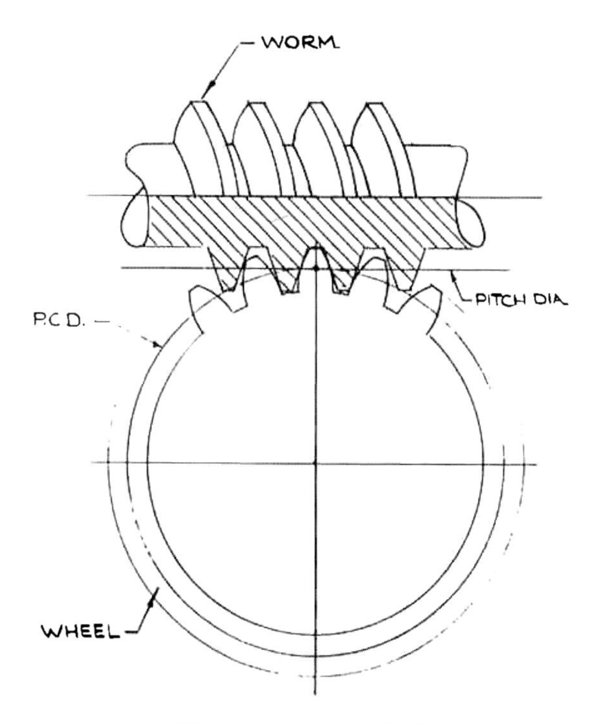

Fig. 46. *A worm and wheel*

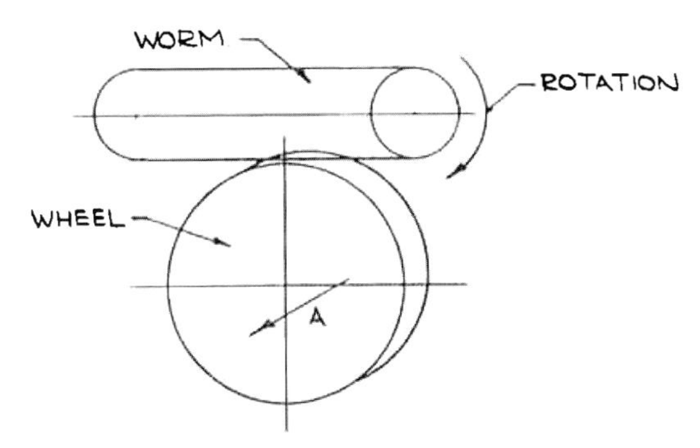

Fig. 47. *Rotation of the worm cylinder will try to move the wheel in the direction of arrow "A" and no turning effort will be produced*

produced or transmitted to the worm. It can be seen from this why it is always necessary to provide a worm with good thrust bearings as most of the effort put into turning the worm will not be transmitted to the wheel but will have to be absorbed by the thrust bearings.

In order to obtain rotary motion, teeth will have to be applied to both worm and wheel and these teeth must be placed at an angle to the axis of the shafts so they can slide against each other. The teeth on the worm are obtained by providing the surface of the worm with a screw thread and the teeth of the worm wheel, which must be of the same form as the screw thread, will have to be angled on the wheel at the thread helix angle otherwise when the correct engagement of the thread and tooth is obtained the two gear shafts will not be at right-angles to each other.

The amount of rotation given to the worm wheel for each revolution of the worm will depend upon the pitch of the screw thread; the coarser the pitch the more movement given to the worm wheel. The coarse pitch will necessitate the teeth on the worm wheel having to be made larger as it is essential that the form of the thread on the worm should be similar to the tooth shape of the worm wheel. Increasing the pitch does not therefore alter the gear ratio between the worm and wheel, the wheel will still only move one tooth for each revolution of the worm. So it can be seen that the gear ratio of the worm and wheel is determined by the number of teeth on the wheel. If the worm wheel has 40 teeth then the worm will have to make 40 revolutions to drive the wheel around once. It can now be seen from the above diagram that the diameter of the worm has no bearing upon the gear ratio as the pitch of the screw thread is not a function of its diameter. Increasing the diameter of the worm will only reduce the helix angle of the thread and its only effect on the wheel will be to reduce the angle of the teeth in relation to the axis of the wheel.

It is possible to alter the gear ratio by putting more than one thread on the worm. For instance, if the pitch of the screw is ⅛" and the wheel has a pitch circle circumference of 3", the wheel would have 24 teeth at ⅛" circular pitch resulting in a gear ratio of 24:1. If the pitch of the screw were to be increased to ¼" and the pitch circumference of the wheel remained the same, there would only be 12 teeth on the wheel and so the gear ratio would now be 12:1. Supposing the number of teeth on the wheel had not been altered but had remained at 24, then at a ¼" pitch the worm would not engage in each tooth of the worm wheel but in alternate teeth and so, using the same gear the ratio has been halved from 24:1 to 12:1.

In practice the result of this arrangement would mean that only a half of the teeth on the wheel were being used and as only the pitch, and not the size of the thread, has been altered, there would be a space between each thread on the worm. Now, if this space were to be used to provide another thread in all respects similar to the original thread, this second thread would be a half turn out of phase with the first thread and if the resultant worm were to be looked at on the end, the threads would have two starting positions, one diametrically opposite the other. This type of thread is referred to as a two-start thread and this second thread will now engage into the gear teeth that the first thread is not using and so all 24 teeth of the wheel are now fully employed. As each thread engages into alternate teeth the gear ratio will be 12:1 and not 24:1. It is possible for a worm thread to have more than two starts—for instance, again using the original 24-tooth worm wheel, if the pitch of the worm were to be increased to ½" it would have 4 starts and the gear ratio would now be 6:1.

When looking at a multi-start thread, measuring the distance between one thread and the next will give the pitch of the thread but this does not mean that this is the amount that the thread will move in making one revolution. Should the thread be two-start then the axial displacement made in one turn of the screw will be twice the pitch; if a three-start then the advance will be three times the pitch, and so on. When discussing multi-start threads it is desirable to use the correct nomenclature. The pitch is the distance between a point on one thread and a similar point on the next thread but the distance moved during one complete revolution of the screw is called the lead. The lead is therefore the pitch multiplied by the number of starts on the thread.

WORM AND WHEEL TOOTH SHAPES

It was seen when discussing spur gears that the shape of the teeth was of great importance; this is true of all forms of gearing and the worm and wheel is no exception. Fig. 48 shows in section a worm and wheel with the section being taken through the center of the worm. The teeth of the worm and wheel can be regarded as the teeth of a rack and pinion and the rules governing the shapes of the teeth of a rack and pinion will apply to the teeth of the worm and wheel. If the involute system is being employed then the screw thread of the worm will have straight sides angled at the pressure angle. The cycloidal system may also be used but the curved shape of the cycloidal rack will have to be reproduced to the shape of the screw thread of the worm. The involute system is invariably used as it is much easier to produce a worm with straight-sided teeth.

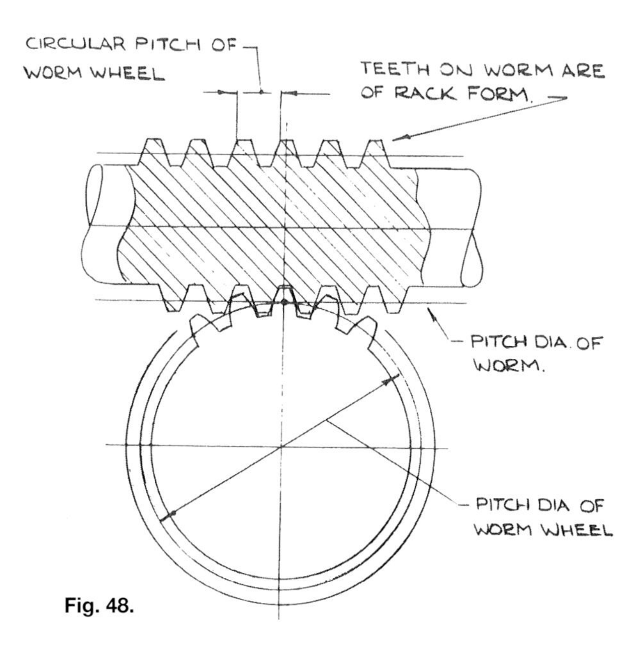

Fig. 48.

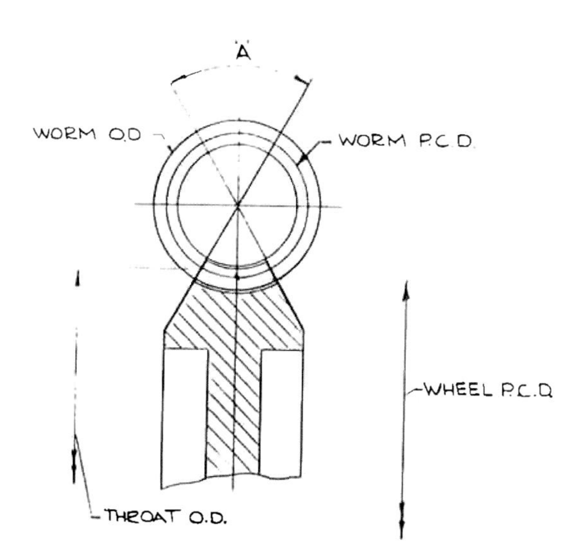

Fig. 49. *The shape of a well-proportioned worm and wheel*

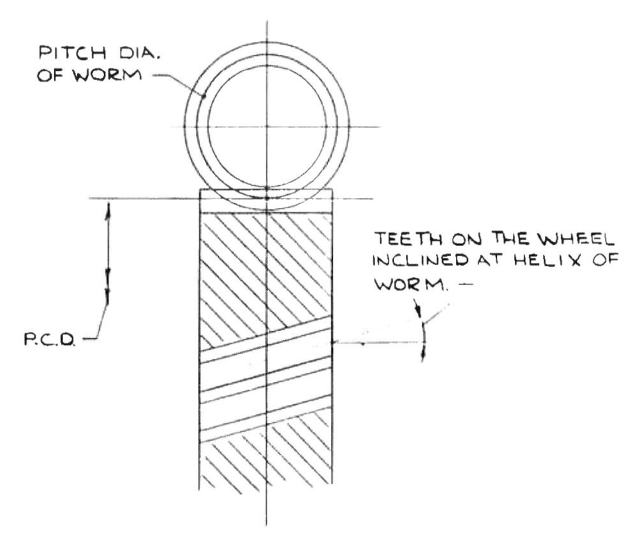

Fig. 50. *Showing a simplified worm wheel with flat topped teeth cut to match the helix angle of the worm*

The worm and wheel will also work satisfactorily if the worm is in the form of an ordinary screw thread as used for nuts and bolts. Of course, the worm wheel must be made with teeth to match the thread chosen. This is very useful for amateur constructors as the most difficult operation in making a worm and wheel in the home workshop is the production of the hob that actually cuts the teeth in the wheel. If a standard thread is used then an ordinary tap that is normally used for forming the thread in a nut can in fact be utilized as a hob; this method will be explained in more detail later.

A correctly shaped worm wheel will have its circumferential face "hollowed out" to suit the radius of the worm, this is shown in Fig. 49. A wheel shaped this way follows the true form of the worm screw thread over the entire width of the worm wheel. In practice the whole width of the wheel is not used, as the corners have a tendency to be weak and also they may not accurately fit the worm. The corners are therefore removed by means of a chamfer. This chamfer is shown in Fig. 49 as an angle of 60° and this angle is usually referred to as the face angle. 60° has been found to give satisfactory results although it is not a mandatory figure and may be varied. A worm and wheel made to this configuration will be capable of long life and high duties and is the preferred profile when the required power

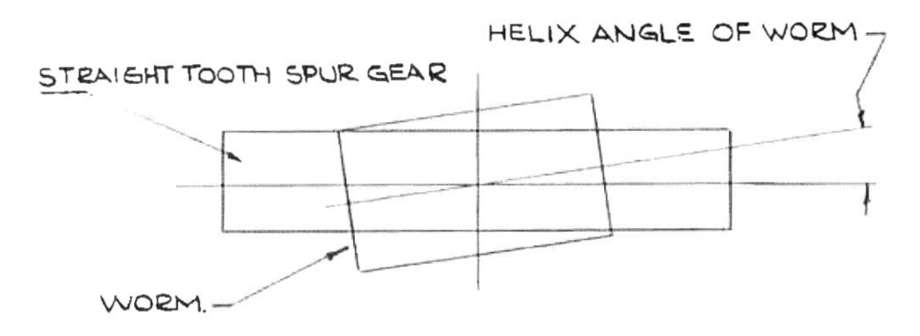

Fig. 51. *Using a straight tooth spur gear as a worm wheel*

transmission is high. Its one disadvantage is that it is not easy to produce, as the wheels have to be individually hobbed and the hob itself is also a complicated shape that has to be made with specialized tooling.

Where the power transmission is not so taxing, or where the duties to be performed are intermittent, the profile of the worm wheel may be simplified. Instead of the curved tooth form the face of the wheel may be flat, similar to that of a spur gear. In order to maintain the axis of the two components at right-angles the teeth of the wheel must be angled, or cut on the slant. This angle, which is part of a helix, must naturally match the helix angle of the worm and this arrangement is shown in Fig. 50. It can be seen from this illustration that the teeth of the wheel do not follow the worm profile and consequently do not possess the same strength and wear characteristics as the profiled or throated worm wheel.

Should it not be an essential feature of the design that the axis of both worm and wheel be at right-angles then it is possible to use an ordinary straight-tooth spur gear as a worm wheel. The axis of the worm must be angled until the thread of the worm is in line with the teeth on the wheel. The angle of inclination will be the helix angle of the worm at its pitch line. This is shown in Fig. 51. There are occasions, particularly for the amateur tool designer and constructor, when this feature can be used to advantage, for instance in the design of a dividing head or rotary table. If the worm is angled away from the table top it will allow the handwheel on the end of the worm to fall below the top surface of the table, and thus allow unrestricted movement for the workpiece.

Worm gears do not generally have a high efficiency, as much of the effort put into the worm shaft is lost in friction and in producing a high end load. For this reason, when worm gears are used for power transmissions a ball thrust bearing has to be used and the whole unit enclosed in an oil bath. This not only provides the necessary constant lubrication but also helps dissipate the heat that the friction causes. The power losses due to friction are diminished as the pitch of the worm increases and so multi-start worms are more efficient than the single start worms.

Unfortunately, the multi-start units generally have a lower gear ratio than the single start version and so the price for a high gear ratio will be a low mechanical efficiency.

The diameter of the worm also affects the overall efficiency: the larger the diameter of the worm the smaller will be its helix angle and so the greater will be the frictional forces, resulting in a lower efficiency. The diameter of the worm must therefore be kept as small as is practical taking into account the loading on both the worm and its bearing. All engineering design is a compromise and a worm gear box is certainly no exception to that rule.

As already stated, worm gearing is not usually reciprocal, that is, although the worm will always drive the worm wheel it is not often that the wheel can be made to drive the worm. If the lead is small so will be the helix angle and any attempt to drive the worm by the wheel would result in the destruction of the unit. Generally speaking, it is only when the lead is increased to the point where it becomes a necessity to use a multi-start worm that it becomes possible to use the wheel as a driver; the greater the lead of the worm the greater will be the likelihood that the gear can be made reversible. The reversibility is of little consequence as it is only in special applications that it is a desirable feature and then the gear ratio required is usually relatively small.

Although the actual proportions of worm gearing may vary a little it is best to keep fairly close to the technical text books on the subject, as these have been based upon years of practical experience. The proportions are usually given in terms of the letter P, where P is the circumferential pitch of the worm wheel teeth. The tooth thickness at the pitch line is quoted as .48P (this allows for a small working clearance between the mating teeth). Height outside pitch line is .3P while the depth below pitch line should be .4P. There is considerable latitude for the length of the worm but about 4P will result in a well-proportioned worm. The face width of the wheel may vary between about 1.5P to 2.5P. It is not essential to keep to these figures but doing so will lead to a well-proportioned unit and a sense of proportion is vital for any engineering designer.

As worm gearing does suffer, unfortunately, from high friction forces that can lead to rapid wear, it is advisable to use suitable materials for the components. The following is intended to act as a guide.

1. A soft steel worm and a cast iron wheel
2. A soft steel worm and a phosphor-bronze wheel
3. A case hardened steel worm and a phosphor-bronze wheel

The ideal is probably a hardened and ground steel worm and wheel although it is unlikely that the amateur constructor will have the necessary facilities to produce the components in this material.

CHAPTER 7

DEFINITIONS AND FORMULAS

Before the actual cutting of gears can commence it is necessary to design the gear required to perform the task in mind and then complete the attendant calculations. As stated at the beginning of this book, the mathematics involved can be kept down to basic arithmetic, so no "non-technical" constructor should have any fears about his ability to successfully undertake the design stage.

From time to time in the preceding pages, certain terms have been used and it may be an advantage to define all the terms collectively and also list the equations that may be needed to set out a train of gears successfully. If all the information is shown together time may be saved as it will not be necessary to look through the body of the book seeking that elusive formula needed to arrive at a particular size.

DEFINITIONS

PITCH CIRCLE

The pitch circles of a pair of gears are those circles co-axial with the gear teeth and bearings that are in peripheral contact and which will roll together without slip. The pitch circles are imaginary smooth rollers or friction discs. The pitch circle is most important as it will represent the actual gear in the layout design stage.

PITCH DIAMETER

The pitch diameter of a gear is the diameter of the pitch circle.

ROOT DIAMETER

The root diameter of a gear is the diameter at the bottom of the tooth spaces. (The root diameter is not usually used in the production of a gear as the diameter that is required in the workshop is the total depth of the cut, usually shown as D+f on commercial gear cutters. If this is applied from the correct outside diameter of the gear blank, then the correct root diameter will be automatically produced.)

CIRCULAR PITCH

The circular pitch is the distance from a point on one tooth to a corresponding point on the next tooth. It is measured around the pitch circle diameter and is not a straight line between the two chosen points.

DIAMETRAL PITCH

The diametral pitch (nearly always referred to as the DP) is the number of teeth per inch of pitch diameter. E.g. if a gear has 40 teeth and the pitch circle diameter is 2" then the teeth are said to be of 20DP.

MODULE

The module of a gear is the reciprocal of the DP but the pitch circle diameter is expressed in millimeters instead of inches. The module is, therefore, the pitch diameter in millimeters divided by the number of teeth in the gear.

THICKNESS OF TOOTH ON PITCH LINE

The thickness of tooth on pitch line is the length of arc on the pitch line between opposite faces of the same tooth.

CHORDAL THICKNESS

The chordal thickness of a tooth is the length of the chord subtended by the tooth thickness arc. It is in fact the straight line between the same two points used in determining the tooth thickness.

CLEARANCE

The clearance is the distance between the top of a tooth and the bottom of its mating space. The clearance is incorporated to prevent contact between the top and bottom of mating teeth and is usually denoted by the symbol "f."

BASE CIRCLE

This is only applicable to involute teeth and is the diameter of the circle from which the involute that determines the tooth profile is generated.

LINE OF ACTION

The line of action of an involute gear is the common tangent to the two base circles which passes through the pitch point of a pair of mating gears.

PATH OF CONTACT

The path of contact of an involute gear is that portion of the line of action at which the tooth contact takes place.

PRESSURE ANGLE

The pressure angle of an involute gear is the acute angle formed between the line of action and the common tangent to the two pitch circles which passes through the pitch point.

ARC OF APPROACH

The arc of approach of a cycloidal gear is that point of the generating circle through which the tooth travels from the time it is in contact with a mating tooth until it is in contact at the line passing between the gear centers.

ARC OF RECESS

The arc of recess of a cycloidal gear is the arc of the generating circle through which the tooth travels from the time it is in contact with its mating tooth on the line passing between the gear centers until the contact with its mating tooth ceases.

THEORETICAL EQUATIONS OR FORMULAS FOR GEARS AND TOOTH PROPORTIONS

Letters used in the following equations are listed below:

PD = Pitch Circle Diameter in inches
PDM = Pitch Circle Diameter in millimeters
CP = Circular Pitch
DP = Diametral Pitch

M = Module
N = Number of teeth in gear
OD = Outside Diameter of Gear (Blank sizes)
T = Thickness of Tooth on Pitch Line
CT = Chordal Thickness
f = Clearance
D+ = Whole Depth of tooth (Depth to set cutter)

Center distance between two gears =

$$\frac{\text{PD of 1st gear + PD of 2nd gear}}{2}$$

Sum of PDs of two mating gears =
2 x Center distance between gears

$$CP = \frac{PD \times \pi}{N} \text{ :or, } \frac{\pi}{DP} \text{ :or, } \frac{OD \times \pi}{N+2} \text{ :or, } 2 \times T$$

$$DP = \frac{N}{PD} \text{ :or, } \frac{N+2}{OD} \text{ :or, } \frac{\pi}{CP} \text{ :or, } \frac{25.4}{M} \text{ :or, } \frac{\pi}{2 \times T}$$

$$M = \frac{PDM}{N} \text{ :or, } \frac{25.4}{DP}$$

$$PD = \frac{N}{DP} \text{ :or, } \frac{N \times CP}{\pi} \text{ :or, } \frac{OD \times N}{N+2}$$

$$OD = \frac{N+2}{DP} \text{ :or, } \frac{CP\,(N+2)}{\pi}$$

$$T = \frac{CP}{2} \text{ :or, } \frac{\pi}{2 \times DP}$$

$$CT = PD \times \text{Sin}\left(\frac{90}{N}\right)$$

$$f = \frac{T}{10}$$

$$D{+}f = \frac{2.157}{DP} \text{ :or, } .6866 \times CP$$

A number of the formulas have been included for the sake of interest only as it is unlikely that the amateur producing his own gears will, in practice, ever have a use for all of them. This is because given the correct form cutter, other than the number of teeth required the only dimension that is needed by the gear cutter is the depth of cut—the form tool will see to the rest! The list may appear to be formidable but in actual fact once a few examples have been followed through no problems should arise. The principles involved can be quickly understood by following a detailed example step-by-step and so we will now look at a typical hypothetical problem.

Supposing it is necessary to produce a pair of gears to reduce the speed of an input shaft from 700rpm to an output shaft speed of 200rpm and that the maximum available space for the gears is 6" x 4". The first step is to find the gear ratio required, and since 700rpm has to be reduced to 200rpm then the required ratio must be in the proportion of 700/200 or 3½:1. This means that the pitch circle diameter of the large gear must be 3½ times greater than the pitch circle diameter of the small gear. Although in theory pinions with very few teeth can be used, in practice there is a limit as with small numbers the teeth become deformed and weak; there is also the problem of providing sufficient material to allow for an adequate bore that is needed to secure the pinion onto its shaft. For general purposes 16 teeth can be considered as the smallest number to place upon a gear although the author prefers, wherever possible, to increase the number to 20.

If, in our example, a 20-tooth pinion is chosen then the mating wheel must have 20 x 3½, or 70, teeth. We have therefore to place a 20- and a 70-tooth gear in a space no greater than 6" x 4" and so the two gears, when running together in mesh, must not measure more than 6" in length and 4" in width. This means that the two pitch circles when added together must be less than 6", because the outside diameters of the gears are greater than their respective PCDs.

In order to determine the maximum size of tooth that can be accommodated in the permitted space the two gear teeth numbers must be added together and one gear of 90-teeth considered. This principle is correct for any number of gears whose centers are on the same straight line; for instance, a train of gears with 20-, 25-, 22- and 23-teeth will have the same overall length as one gear of 90-teeth.

Referring to the formulas we see that $OD = {}^{N+2}/_{DP}$. Therefore, we apply our figures to this equation and we can determine the maximum DP that can be accommodated in the permitted space, viz: $6 = \frac{90+2}{DP}$ or

$DP = \frac{92}{6}$, which is 15.3. The nearest standard DP that could be used is 16, which means that the maximum size of gear that can be accommodated in the 6" length restriction is 16DP. Of course, any smaller tooth size could be used as this would lead to smaller gears. By using the same equation again the outside diameter of a 70-tooth 16DP gear can be calculated, viz: $OD = \frac{70+2}{16}$ which is 4½". This is greater than the 4" permitted and so we see that the limiting factor is not the 6" length but the 4" width.

By using the same formula once again the maximum tooth size for the 70-tooth gear can be found, $4 = \frac{72}{DP}$ or 18DP. This would completely occupy the 4" space but as it would be an advantage to have a working clearance 20DP would seem to be the largest practical tooth size that can be used. We can now obtain some definite figures for the two gears. The pinion would be 20-teeth—20DP; from the list of formulas we see that the $PD = {}^{N}/_{DP}$, in this case $^{20}/_{20}$, resulting in a pitch circle diameter of 1". The outside diameter of the gear blank would be $OD = {}^{N+2}/_{DP}$, or $^{22}/_{20}$, resulting in a figure of 1.1". By applying the above to the 70-tooth wheel we get 70 teeth, 20DP, PCD 3.5" and outside diameter of blank 3.6".

The center distance of the two gears can be obtained by adding together the two pitch circle diameters and dividing by 2—hence, $\frac{3.5+1}{2} = \frac{4\frac{1}{2}}{2}$ or 2¼". The overall length of the two gears in mesh will give a figure of 4.6", well within the original 6" allowed but, as was shown, the limiting factor was the width restriction.

In an engineering design office, the calculations would be considerably extended to take into account the stress loading of the gear teeth, etc., and from these figures the minimum size of gear would be chosen, but one that would be capable of transmitting the power needed. Any serious engineering student should follow this line of procedure. In model engineering and work generally undertaken in the home workshop, not only gears but most components are not highly stressed parts so if the designer or constructor has a sense of proportion and also applies the old adage that "if it looks right it most likely is all right," then in the majority of cases satisfactory results will be achieved.

CHAPTER 8

DIVIDING HEADS

One of the main components in the tooling system required for cutting gears is a method of indexing the teeth correctly. There is little point in being able to cut the gear teeth to the profile needed to give a smooth and even power transmission if the gear teeth themselves are not correctly spaced. The tool used to give the correct spacing is the dividing head. A dividing head can be a very simple device or it may be a sophisticated one but basically they are all the same. They consist of a headstock into which a spindle is mounted and one end of the spindle will have some means of securely holding the workpiece while at the other end there will be a method which permits the spindle to be moved through a definite and predetermined angle and then locked in that position.

Dividing heads can be separated into two types, the simple direct division head and

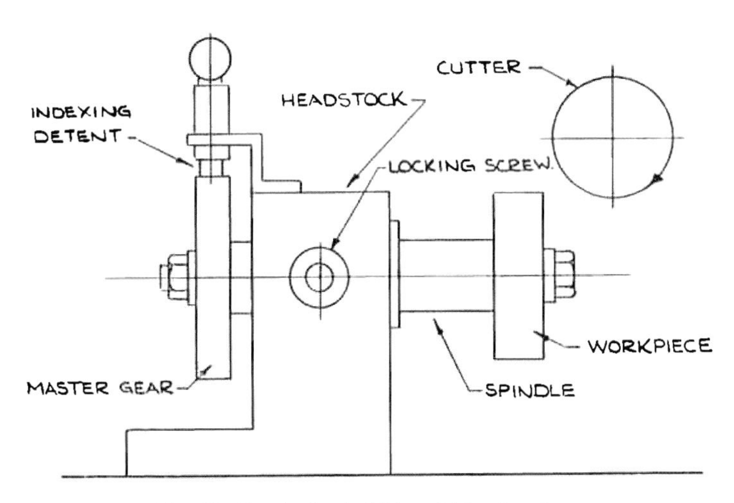

Fig. 52. *Simple direct division dividing head*

the more complex worm and wheel head. A simple direct division head is shown in Fig. 52. Here we have a headstock that consists of an angle, one leg of which is bolted onto the work-table of the milling machine or lathe, the other—or vertical—leg carrying the spindle. The gear to be cut is secured to the spindle by means of a clamping nut and on the other end, again securely mounted, is a master gear and detent. The master gear is used to copy or space the teeth into the workpiece. The detent, which consists of a spring-loaded plunger with a conical or V-shaped point, drops into the gap between two teeth of the master gear, thus positively locating the spindle which is then locked to prevent any load being placed upon the detent while a tooth is cut on the workpiece. After cutting, the spindle is then freed and moved until the detent drops into the next gap, this process being repeated until all the teeth have been cut.

One drawback with this method is that there must be a suitable gear available to use as a master gear; however, the master gear need not have the same tooth size as that required on the workpiece. A 20DP gear could be used as a master, yet the gear being cut could be 32DP, or any other size within reason. Nor is it necessary for the master gear to have the same number of teeth as the workpiece, but there must be some common factor between them. For instance, if it is required to cut a 20-tooth gear, the master could have 20 teeth and the detent would then have to be engaged into every tooth space, but a master may be used having 40 teeth in which case the detent would then be engaged into alternate teeth, or a 60-tooth gear could be used to produce 20 teeth and indexed one tooth in three.

Lathe changewheels make good master gears and as these come in a set a wide choice is available. The 60-tooth gear is very useful as this will give divisions of 2, 3, 4, 5, 6, 10, 12, 15, 20 or 30. Even though a large selection of changewheels may be to hand, there are severe limitations to the numbers that they can index as this type of gear is usually supplied in increments of five teeth only. If the detent is made so that it can index over the teeth as well as between them it will double the number from which factors can be obtained (the 60-tooth gear has 120 divisions and so on) but the increase in the range will be negligible.

Another drawback to the direct division head is that the accuracy of the workpiece will depend upon the accuracy of the master gear: any errors in pitch that are present in the master will be passed on to the workpiece. This drawback is more theoretical than practical as most lathe changewheels are now produced by one of the generating processes, such as hobbing, resulting in a gear far more accurate than can be achieved by the "one tooth at a time" method used by the amateur.

The direct division head can be made more sophisticated than the example outlined and Fig. 53 shows such a unit. This is provided with a screwed mandrel nose and also a Morse taper. The spindle locking is obtained by clamping a split bearing instead of the clamp or pinch-screws. This head also has a secondary detent arm so that the gears may be compounded in order to increase the range of divisions available. It is, however, still a direct division head and as such has its limitations. One advantage the direct head possesses, other than its simplicity, is that within its operating range indexing is quicker

Fig. 53. *The M.E.S. Direct Division Dividing Head. This is only supplied as a kit of parts suitable for home construction*

than the worm and wheel type and there is less chance of making an error.

WORM AND WHEEL TYPE HEAD

The worm and wheel dividing head is basically a direct division head with one important addition. The master gear mounted on the spindle or mandrel is a fixture and is not changeable. The gear is in fact a worm wheel and is driven around by means of a second shaft carrying a worm. The worm shaft is extended upwards and firmly secured to its outer end is a crank-handle, thus providing a means of rotating the worm. The worm is a single start so that one rotation of the handle will advance the worm wheel one tooth.

It is usual general engineering practice for the ratio of the worm and wheel to be 40:1. The Myford dividing head, Fig. 54, is, in fact, a 60:1 ratio and the author has found this an advantage for the model engineer and home workshops as the factors of 60 are more commonly used than the factors of 40. One turn of the handle of the Myford head will rotate the spindle and workpiece 1/60th of a turn, 10 turns of the handle will give a rotation ⅙ turn and so on. By moving the handle a part of a turn it is possible to rotate the spindle through any desired angle and therefore an infinite range of divisions is theoretically possible.

Some form of position indicator to register the amount of turn is needed and this is called the division plate. The division plate is

Fig. 54. *The Myford Worm and Wheel Dividing Head with one of the additional Division Plates*

mounted concentric to the worm but is fixed to the head itself and does not rotate with the worm. A series of holes is drilled into the plate concentric to the bore and the handle is provided with a spring-loaded plunger so that the locating pin on the plunger can be placed into any hole as desired. In order to get maximum use out of one division plate a set or series of holes is drilled each on a different pitch circle diameter and the arm that houses the plunger is slotted so that the plunger can be aligned over the requisite circle of holes. The Myford head is supplied with two division plates that contain in all 15 sets of holes which enable most divisions between 1 and 100 to be obtained. The few that are not available are mostly prime numbers and are unlikely to be needed but if they are, two more plates are available as an extra and they will fill in the few missing numbers. The chart supplied with the dividing head lists all the combinations of divisions available and quotes the number of complete turns required for each division, plus, where necessary, the number of holes needed in a specific division plate.

Counting the holes can become confusing; if 28 holes were required in a 49-hole division plate then this counting would have to be performed for every division and so an error could easily arise. To help overcome this problem an aid is provided which consists of two blades that can be rotated around the worm shaft and just above the division plate. The blades can be moved independently of each other and then locked together at any desired position so that they then move as a unit. They can therefore be set to embrace the number of holes required—28 in the 49-hole in the above example—and so indicate the hole the plunger has to enter to achieve the correct spacing. After each movement of the crank-handle the blades are rotated as a pair until one of the blades makes contact and rests against the plunger, the next hole for the plunger then being indicated by the other blade.

The author never likes to have to rely solely on tables or data charts for information as they usually get torn or covered with oil and dirt, making them difficult to read. It is always prudent to be able to work things out for oneself. There is nothing magical about determining the number of holes or which circle to use in the dividing plate, as the following example shows.

Supposing that one wishes to cut a gear having 33 teeth using a dividing head with a worm and wheel ratio of 60:1. The first step is to divide the number of divisions required into the worm and wheel ratio—in this case 60 divided by 33. This gives an answer of $1\frac{27}{33}$, which indicates that each division is one complete turn of the handle plus a further addition of $\frac{27}{33}$ of a turn. The $\frac{27}{33}$ of a turn can be achieved by using 27 holes in a 33-hole circle of the division plate.

If a 33-hole circle is available then the problem is solved but it is unlikely that a 33-hole circle will be on the division plate, so the next step is to look at the fraction $\frac{27}{33}$ and see if it can be factorized. And the answer is, yes it can, as both top and bottom are divisible by 3, resulting in a new fraction of the same value $\frac{9}{11}$. If the division plate does not have a 33-hole circle then it certainly will not have an 11-hole one but it is most likely to have a circle that is a multiple of 11 and this will in all probability be a 77-hole circle. If we now return to our $\frac{9}{11}$ fraction and multiply both top and bottom by 7 we will arrive at $\frac{63}{77}$. The division for producing a 33-tooth gear is therefore one complete turn of the handle plus 63 holes on a 77-hole circle of the division plate. Once this example is understood it is a simple matter to substitute any division as required.

The occasion may arise when the division plate does not possess a circle with the requisite number of holes. This is a most unlikely occurrence as the plates are carefully planned to give a very wide range of divisions but, if a makeshift division plate has to be made do not worry about positioning the holes to very fine limits, as any error in the division plate is not passed on directly to the workpiece but is divided by the worm wheel ratio and so the actual error becomes very tiny indeed.

CHAPTER 9

CUTTING SPUR GEARS

A friend of the author was so pleased when he got his new lathe working for the first time that he felt he must show it to someone. The only person on hand was his wife and although he knew she had no idea what a lathe was or what it did, he invited her into the workshop for a demonstration. He put on a cut, set the machine in motion and stood back with pride, "What do you think of that?" he asked. His wife thought for a moment then said to him "Very nice, darling, but tell me, what do you do with those curly things when you have made them?" Now, that lady had observed something that many of us are apt to forget, and that is no matter what machine we are using, be it a lathe, milling machine, drilling machine, shaping machine or any other machine tool, they all produce the same thing—swarf, and it is the bits that are left behind that matter. Gears are no exception and although we have gone to great trouble to determine the shape of the teeth we do not cut **them**, only the spaces between them, and so a gear cutter is a form cutter—that form being the space between two adjacent teeth.

It was seen when discussing the involute shape that the base circle from which the involute is formed is related to the PCD of the gear and so, as the number of teeth in the gear alters, so does the diameter of the base circle and this is reflected in the shape of the involute curve. From this it follows that the shape of the tooth for any given DP varies according to the number of teeth in the gear. It would appear from this that a different cutter is needed to suit every size of gear that may be required and this would represent a lot of cutters!

The involute curve changes quite markedly between gears that have only a small number of teeth but the rate of change slows down and becomes less apparent as the gears get larger, and once the number of teeth passes the 120 mark then the change in shape is negligible. A slight discrepancy in tooth shape can be tolerated in general purpose gears without any noticeable effect in the gears' performance. These two facts are recognized by the gear cutter manufacturers and so it has been possible to reduce the number of cutters required for each DP to eight. Only a few gears can be cut with a cutter from the lower end of the range but as the gears get larger, with a corresponding rise in the number of teeth, then the range of the cutter also increases. In fact the No. 1 cutter will produce any gear possessing over 134 teeth and this includes the rack.

All disc-type gear cutters that cut only one tooth at a time are usually referred to as Brown & Sharpe cutters, this being the name of the company that developed this type of cutter many years ago. They are now made by cutter manufacturers all over the world but all meet the international standard based on the work done by Brown & Sharpe.

Each cutter in the range is given a number which will clearly be marked on the cutter together with the number of teeth for which that particular cutter is designed. The table of cutters is listed below:

Cutter No. 8 will cut gears of 12 and 13 teeth
Cutter No. 7 will cut gears of 14–16 teeth
Cutter No. 6 will cut gears of 17–20 teeth
Cutter No. 5 will cut gears of 21–25 teeth
Cutter No. 4 will cut gears of 26–34 teeth
Cutter No. 3 will cut gears of 35–54 teeth
Cutter No. 2 will cut gears of 55–134 teeth
Cutter No. 1 will cut gears of 135 teeth to a rack.

Once having selected the correct cutter the next problem is to set the cutter so that it will produce the tooth to the correct depth. The datum for setting the depth of cut must be the outside diameter of the gear blank as there is no other datum from which to measure. It was shown in Chapter 7 that the diameter of the gear blank must be $N+2/DP$. From the practical point of view the outside diameter of the finished gear is not highly critical and a reasonable machining tolerance, particularly in the minus direction, is permissible, but what is critical is the size of the PCD and this will be influenced by the depth of cut. It will therefore be prudent to measure the outside diameter of the blank and check it against the theoretical size to see if any discrepancy has arisen. If it has then this must be taken into account when setting the cutter to depth. If, for example, the gear to be cut is a 40-tooth 20DP then the outside diameter of the blank should be $42/20$ or 2.100", but if, on measuring, it is found to be 2.090" then this .010" error on diameter must be taken into account and the depth of cut set on the machine reduced from the normal depth by .005". In this instance, instead of setting to .108" the amended cutting would be .103".

The method used for setting the depth is to bring the blank up to the cutter until it just touches, set the micrometer dial of the relevant machine-slide to zero and then, using the micrometer dial, put on the correct cut and securely lock the slide in that position. It may not be particularly easy to determine just when the cutter is touching the workpiece and if one goes too deep the datum is lost. This would mean indexing to the next tooth and trying again! The author has found that it is a great help to place a piece of thin wet tissue paper on the blank then, with the machine running, carefully bring the cutter and workpiece together until the cutter removes the paper. When this happens the setting can be regarded as being correct.

The depth of cut will normally be determined by the DP or CP and the formulas for calculating these is shown in Chapter 7. The figure is also usually quoted on the actual cutter although it does not say depth of cut as such but is prefixed by the term D+f. In order to save time calculating the depth each time it is required the following tables have been compiled. The depth is shown in both imperial and metric sizes, thus enabling the depth to be set directly depending on the graduations of the micrometer dials.

DP	Depth of Tooth in Inches	Depth of Tooth in mm
6	.360	9.13
8	.270	6.85
10	.216	5.48
12	.180	4.56
14	.154	3.91
16	.135	3.42
18	.120	3.04
20	.108	2.74
22	.098	2.49
24	.090	2.28

DP	Depth of Tooth in Inches	Depth of Tooth in mm
26	.083	2.11
28	.077	1.96
30	.072	1.83
32	.067	1.71
36	.060	1.52
40	.054	1.37
48	.045	1.14
64	.034	.86
72	.030	.76
80	.027	.68

Module	Depth of Tooth in Inches	Depth of Tooth in mm
1	.085	2.16
0.8	.070	1.73
0.7	.059	1.51
0.5	.042	1.08
0.4	.034	.86
0.3	.025	.65

Module	Depth of Tooth in Inches	Depth of Tooth in mm
4	.340	8.63
3	.255	6.47
2.5	.212	5.39
2.0	.170	4.31
1.5	.127	3.24
1.25	.106	2.70

Always try to cut to the correct depth with one cut; nibbling away with a succession of small cuts is not recommended. Commercial gear cutters are expensive items and naturally the amateur wants them to last as long as possible between re-grinds. It is a fallacy to think that you are being kind to the cutters by making numerous light passes. Set the cutter to the correct depth and regulate the feed to suit the machine, naturally a heavy robust machine will be capable of a faster feed rate than a small lighter one. When cutting large teeth such as those found on traction engine models, the equipment available may not be man enough to produce the tooth at one pass in which case it may be an advantage to "gash" the teeth first with either a slitting saw or a side-and-face cutter. This will remove the bulk of the material and lighten the load on the machine when the form cutter is used.

In industry time is an important factor and speeds and feeds are designed to give the maximum swarf removal rate commensurate with overall costs. Production machines are specially built to reduce machining times to as little as possible so it follows that speeds and feeds published in industrial text books are of little use to the amateur as the machine tools at his disposal may be no more than a lathe and a milling machine and both of these will be flimsy when compared to the production machines. Very few machines in the home workshop will be equipped with a full flow cooling system, in fact the majority will have no cooling system at all other than a "can and brush" or possibly a drip can.

The table of suggested RPMs has taken all this into consideration but, even so, regard the figures as a guide only and do not be tempted to run the cutter too quickly. It is most unlikely that the speed range of the machine available will correspond to the figures quoted in the table, this is of little consequence, try the speed nearest to the figure quoted and see what happens! This may sound rather haphazard but there are so many unknown factors in the amateur set-up that quoting firm optimum speeds is, to say the least, an exercise fraught with danger.

Most commercial cutters of the Brown & Sharpe type that the amateur is likely to have will be around 2⅜" dia. with a 1" dia. bore, as this is a standard size. The cutters will be high speed steel correctly heat-treated and ground and will, naturally, be capable of a heavier work load than a home-made cutter. Home-made cutters will be of carbon steel, such as drill rod, with the heat treatment carried out without the advantages of temperature controlled equipment. The first column in the table is for the commercial cutters while the remainder are for differing diameters of home-made cutters.

The back garden workshop will not normally possess any specialist machinery so the gear cutting will have to be performed on either a lathe or a milling machine. If a miller is available then this will normally be first choice but if not, then the lathe can be successfully adapted for the task.

A gear cutting device has two main elements, one is to hold and index the blank and the other one is to provide a means of holding and rotating the cutter at the correct speed. The lathe can be adapted in two ways to fulfill the above requirement. Firstly, the cutter can be held in some holder or arbor secured to the lathe mandrel, with the gear blank together with its dividing head secured to the cross-slide of the machine.

In the second method the roles are reversed and the blank is fastened to the lathe mandrel while the cutter, with some form of auxiliary drive, is attached to the lathe cross-slide. This second method used to find considerable favor but its popularity has very much declined in recent years owing to the changes in the design of the lathe itself. Nearly all modern small lathes are self-contained units and not only are they motorized but even the drive belts are now completely guarded, making access to the belts and pulley difficult,

Material	Brown & Sharpe Commercial HSS	Home-made Drill Rod Cutter		
	2⅜" dia.	1¼" dia.	1½" dia.	2" dia.
Steel	60rpm	75rpm	62rpm	48rpm
Brass	100rpm	120rpm	100rpm	75rpm
Light Alloy	140rpm	170rpm	140rpm	100rpm

particularly when the lathe is running. The reason for this, of course, is to safeguard the operator and protect him from the belt, pulleys and gears which are potentially dangerous.

In the early days of the small model engineering lathe, particularly between the wars and even up to the fifties, many lathes were driven by flat belts from an overhead counter- or line-shaft. This shaft would be driven by the workshop's only motor and often ran the entire length of the workshop to be used as a power take-off for all the machinery that the workshop possessed. The overhead counter-shaft was a handy place for supplying power to auxiliary cutting heads that could be mounted onto the lathe bed or cross-slide, and many attachments, not only for gear cutting but also for milling in general, were powered in this way. The author's first workshop was arranged in this fashion and was a cumbersome affair with open flat belts and pulleys whirling around in all directions and standing in the middle of all this activity could be, by modern standards, a little alarming. However, the system had its advantages, particularly when it came to auxiliary drives, although the most vivid memories are of continually trying to solve the problem of joining flat belts in such a way that they would not clatter, break or give the hand a nasty cut each time the speed was changed which, unless the whole workshop was immobilized, had to be done while the machine and its belts were running.

The advent of the Myford ML7, and similar types of lathe, in the late forties brought about a complete revolution not only in the author's workshop but in home workshops in general and now almost all machine tools are self-contained and so very few overhead drives remain. As there are no handy rotating shafts to provide a supply of power to a cutter mounted on the lathe cross-slide and as all self-contained motorized units can be heavy bulky lumps to adorn a cross-slide, this method of cutting gears in the lathe is not now often used. On the occasions that it is favored it is generally confined to small, light gears, such as instrument or clock gears, where the cutter speed can be high and so eliminate the need for a reduction gear box in the drive to the cutter.

The first method quoted, where the cutter is driven by the lathe mandrel, has several features to commend it. The whole speed range of the lathe is available to the cutter and speeds can be chosen to suit the size and type of cutter as well as the nature of the gear blank material. The full power of the lathe is also available so that speeds and feeds can be regulated to suit the work rather than to suit the power available from a small auxiliary motor. There is also much greater inherent rigidity with this arrangement as the cutter arbor is rotating in the lathe mandrel bearings which are much more robust than those that could be fitted into a convenient drive mounted on the cross-slide. Fig. 55 shows diagrammatically how the various units are arranged for cutting a gear.

The dividing head is mounted on a vertical slide in such a way as to ensure that the spindle is parallel and in line with the lathe cross-slide. If both the dividing head and the vertical slide are Myford units then the correct alignment will automatically be ensured as the dividing head is fitted with a locating key which registers in the T-slots of the vertical slide. If other types of vertical slides and dividing heads are being used, check the alignment very carefully because if the spindle of the dividing head is not parallel with the cross-slide then neither will be the arbor carrying the gear blank and the resultant gear teeth will be deeper at one end of the gear than at the other. Likewise, if the spindle and arbor are not in line with the cross-slide or square with the lathe bed, the teeth will not be parallel to the bore of the gear but at an angle to it. Both these faults must be avoided as either could seriously affect the performance of the finished gear.

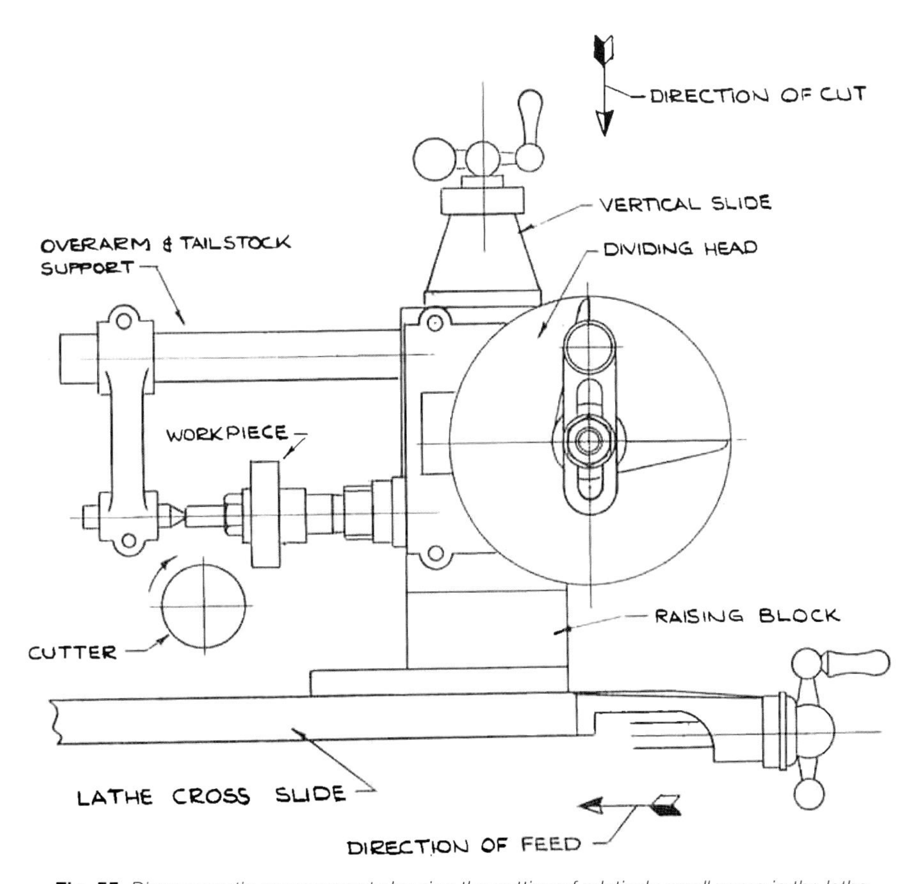

Fig. 55. *Diagrammatic arrangement showing the cutting of relatively small gears in the lathe*

Fig. 56. *Showing the Myford Dividing Head fitted onto a vertical slide and mounted on a raising block*

One disadvantage of using this particular arrangement to cut gears is that unless the vertical slide is abnormally long it is not possible to lift the center line of the dividing head high enough to allow a moderate size gear to be cut. This can be overcome to some extent by the use of a raising block, the block being positioned between the cross-slide and the vertical slide. Ideally the raising block should perform a dual role, that of lifting the dividing head and also displacing it towards the tailstock. This will have the effect of moving the dividing head away from the close proximity of the headstock, thus providing more space for the operator to crank the handle. The photographs Figs. 56 and 57 illustrate this point and also show the complete rig.

The cutter is, of course, mounted on an arbor and is driven by the lathe mandrel. The arbor will have to be lengthy to provide a working space for the dividing head and it will also have to be supported at the outer end by the tailstock center. The arbor and mandrel upon which the gear blank is secured must be firmly located and held in the dividing head and it is also an advantage to provide an outer support for this shaft; the illustration shows how this is accomplished by using an overarm which is a feature of this particular dividing head.

The author always prefers both cutter and workpiece arbors to be provided with a Morse taper shank so that they may be fitted directly into their respective spindle noses. Although this may add a complication to their manufacture the extra work is well worthwhile. They can be used time and time again without any need to set them up to run truly; they also eliminate the need for any further holding device such as a chuck, which will not only get in the way but will reduce the overall rigidity. It is also recommended that both the arbors are secured by means of drawbolts passing through the hollow mandrels of the lathe and dividing head.

In the rig under discussion the cutter will revolve towards the dividing head and so the forces created by the cutting action will be passed on to the thrust bearings of the dividing head, which have been designed to absorb them. It would be bad practice to cut towards the tail support as this is not intended to take the thrust loads. In view of the above, it may seem unnecessary to use a drawbolt but not only are thrust loads created by the cutting action, so also are shock loads and these are the ideal loads to withdraw "stuck" tapers. It is not a good idea to run the cutter between centers as the load on the cutter, particularly when starting a cut, is intermittent and the rapid change between load and no-load will cause the cutter to rotate in a series of jerks and give rise to unpleasant vibrations. The load conditions are still the same when the

Fig. 57. *Another view of the arrangement shown in Fig. 56. Even by using a raising block, height limitations restrict the size of gear that can be cut*

cutter is mounted in the Morse taper of the lathe spindle but the mass of the whole lathe drive has a dampening effect and the cutter rotates more evenly.

The cut is applied by means of the vertical slide feed-screw handwheel and the feed to the cutter is supplied by the cross-slide screw. If the lathe is fitted with a power cross-feed this may be used providing that the rate of feed is suitable. It is important that the center line of the form tooth shape must be coincident with the center line of the arbor onto which the gear blank is secured. If this condition is not met the teeth produced on the gear will not be truly radial but will lean over to one side with disastrous results to the correct meshing of the finished gear. There are ways of obtaining the correct setting by measurement and one method is to remove the gear blank from the arbor and then by careful manipulation of the lathe handwheel position the arbor until its side just touches the side of the cutter and then note the reading on the lead screw handwheel. If, after lifting the arbor upwards from the cutter the lathe carriage is moved a distance equal to a half of the arbor diameter plus a half of the cutter width, the center of the arbor should be directly over the center line of the cutter. The lead screw's micrometer dial should be used to obtain the setting as accurately as possible.

This method, and variations of it, have been tried by the author but he has found that the quickest and most reliable method is to position the cutter by eye. To do this remove the arbor from the dividing head and replace it with a lathe center. Then line up this center with the center line of one of the cutter's teeth. Naturally, the cutter line is not indicated on the tooth and has to be estimated but by using a good magnifying glass the correct setting can be obtained within fine limits. After setting, the lathe saddle is securely locked in place, the center removed and the arbor replaced. The photograph Fig. 58 shows this centering being carried out.

Before starting to cut the first tooth, set the length of the crank handle on the dividing head so that the detent pin engages into the correct circle of holes. If the number of teeth being cut will divide directly into the number of teeth on the worm wheel then any circle of holes may be used because only one hole will be used, the division always being a complete number of turns of the crank handle. No matter how many circles of holes there are in the division plate the first or datum hole of each circle will be on the same radial line.

It is recommended that the datum hole is used as a starting point, although a start can be made from any hole; however, it is a good idea to use an easily recognized hole as a starting point because after one complete revolution of the dividing head the detent pin should finish in the starting hole. If it does then all is well, if not then an error has been made somewhere in the divisions. It is also an advantage to make a habit of always rotating the crank handle in the same direction as this will eliminate any accidental backlash errors. If the chosen direction is clockwise then approach the first datum hole in a clockwise direction and if an over-run occurs go back about a quarter turn and try again; never go back to the hole. This also applies to any overshoot during the complete indexing process. Once the correct setting has been established lock the dividing head by means of the pinch-screws provided, lock the lathe saddle in position and also the vertical slide. The only free movement during the actual cutting should be the slide that is feeding the cut.

Fig. 58. *Positioning the cutter by using a lathe center as a temporary arbor replacement*

As mentioned earlier, there is a limit to the size of gear that can be cut by this arrangement owing to the height restrictions imposed by the vertical slide. Even with a raising block only moderate sized gears can be accommodated. When determining the largest gear that can be cut the governing factor is the distance between the lathe center and the center line of the dividing head when the dividing head is set as high as the vertical slide will allow. This distance has to accommodate not only the radius of the gear blank but also the radius of the actual cutter. The two radii together must not exceed the distance between the two center lines. A smaller diameter cutter will allow a larger gear to be cut, a point worth considering if the cutters are to be "home made." The cutter shown in Figs. 56, 57 and 58 is a standard Brown & Sharpe cutter of 2⅜" dia. and using this the 25-tooth, 20DP gear being cut is well within the range of the rig, but when the rig's limit has been reached some other arrangement of tooling is needed and this is shown in Fig. 59.

The equipment is basically the same as was used in the previous method except that this time the vertical slide must be of the swiveling type, whereas before a fixed 90° type was actually used. A swiveling type would have been perfectly satisfactory but it would have had to be locked into the 90° position. The cutter is mounted in exactly the same way as in the previous method, in fact the same arbor can be used for both arrangements. The swiveling vertical slide is attached to the raising block at an angle as shown in Fig. 59 with the feedscrew handwheel pointing downwards, which has the effect of elevating the spindle nose of the dividing head and providing more headroom. The angle of inclination is not important but it could be influenced by the size of the gear to be cut, a larger gear calling for a greater angle than a relatively smaller one.

The vertical slide has three T-slots and any one of them may be chosen as a mounting point for the dividing head. Both this and the actual position along the T-slot chosen as the mounting point influence the size of gear that can be cut. Once again it is imperative to set the center line of the dividing head parallel to the center line of the vertical slide, otherwise the tooth depth across the face of the gear will vary throughout its length. When centralizing the cutter to the workpiece it may not be possible to lower the dividing head far enough to allow a center to meet the cutter. If this is the case then swing the swiveling vertical slide over until contact can be made. Make the adjustments and then firmly lock the lathe saddle in position. The vertical slide can then be swung back to get the dividing head to the required angle. This will not alter the setting as the movement will be in the plane of the setting. The photograph Fig. 61 shows this centralizing being performed.

The methods of applying both the depth of cut and the feed are totally different from those used previously. The feed in this case is applied by the handle or handwheel of the vertical slide and this is made possible

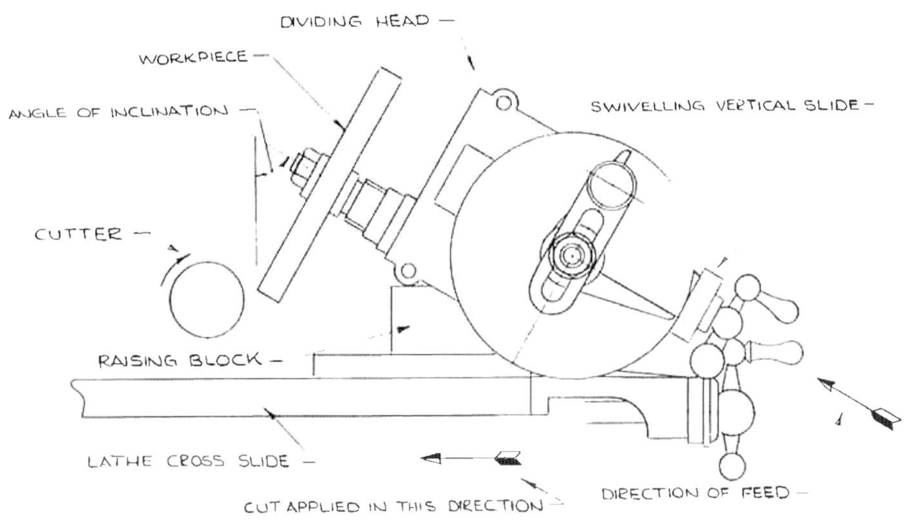

Fig. 59. *Diagrammatic arrangement showing the rig for cutting the larger type of gear in the lathe*

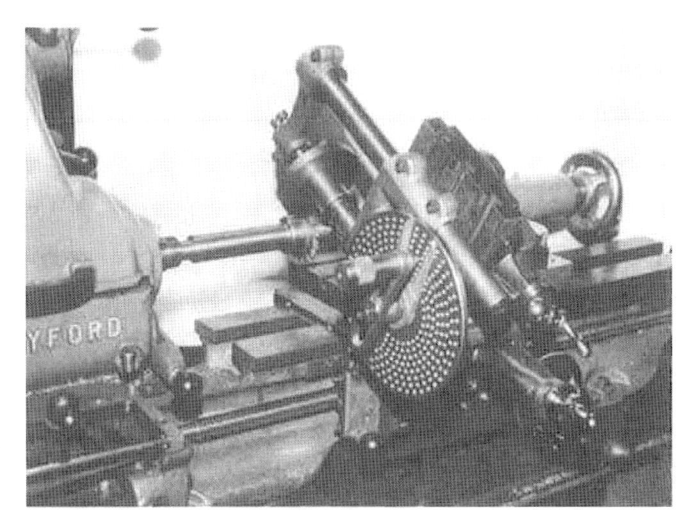

Fig. 60. *This shows how the swiveling vertical slide is used to provide more headroom for the accommodation of larger gears*

Fig. 61. *Centralizing the cutter. The vertical slide has been set at a low angle to allow the center access to the cutter*

because the spindle of the dividing head is in a plane parallel to the slideway of the vertical slide. It does, of course, prohibit the use of a power feed but this is unimportant and should not present any problems. There are times, in fact, when feeding by hand is an advantage because it is possible to get the "feel" of the cutter and apply the feed accordingly.

Obtaining the correct depth of cut does present a problem as there is no means of feeding the cut at right-angles to the axial center line of the workpiece. The feed has to be applied by the cross-slide screw which is at an angle to the movement desired. It has already been explained how important it is to cut the teeth to the correct depth, so some means must be found of determining the depth of cut. It may be possible to obtain this by cutting a tooth deliberately shallow and then testing the depth of cut by means of

Fig. 62. *This large gear precludes the use of a tail support. A short arbor is used to obtain maximum rigidity*

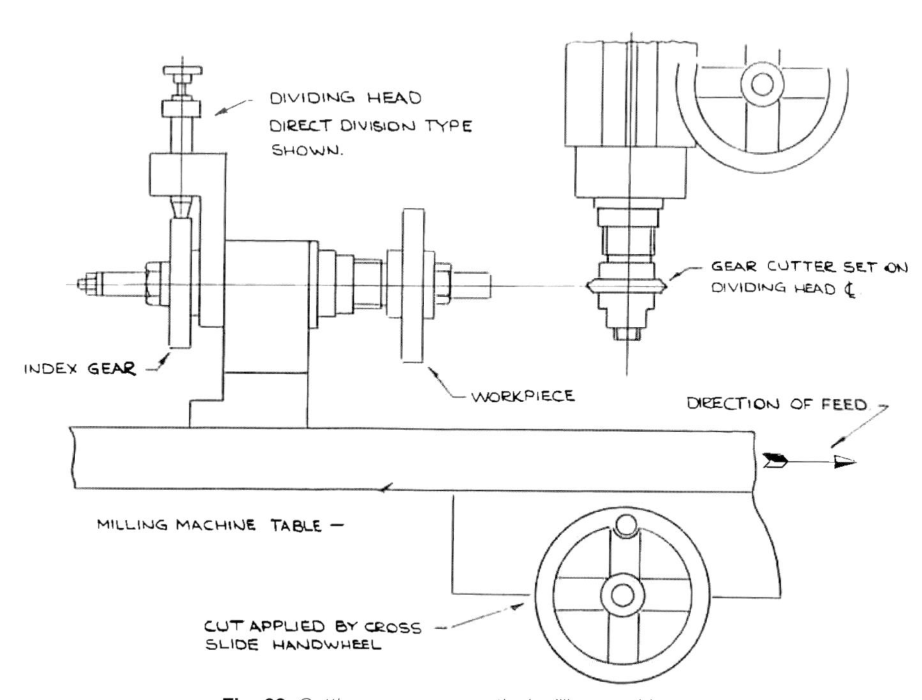

Fig. 63. *Cutting gears on a vertical milling machine*

a depth gauge, adding a little to the cut and then repeating the operation until the correct depth is reached. All the teeth can then be cut to this setting.

The author, however, is not keen on this method as obtaining an accurate depth reading is somewhat haphazard. This is a case where a knowledge of elementary trigonometry is a good tool to have in our possession. If the angle of inclination of the dividing head is known then it is a simple calculation to determine the distance the cross-slide has to move to arrive at the required depth of cut. If the desired tooth depth is divided by the cosine of the angle the dividing head spindle makes with the vertical plane, the result will be the distance the cross-slide needs moving. A 60° angle is a convenient angle to use because the cosine of 60° is .5, so the amount the cross-slide is moved is simply twice the depth of cut required. Unfortunately, it may not always be convenient to set the vertical slide to 60°. The angle of inclination chosen will most likely fall somewhere between 30° and 65° and so the following table has been compiled in increments of 5°. The figures shown are the amount the cross-slide screw needs advancing after contact has been established between the cutter and the workpiece.

The photograph Fig. 60 shows a 63-tooth gear being cut; the gear arbor is shown with a tail support, this support being provided by the overarm in the dividing head. With the dividing head shown the overarm support can be used on gears up to about 5¾" dia.; for gears larger than this the overarm must be removed, otherwise it would foul the gear. This is a pity because as the gears get larger the greater is the benefit derived from the support. However, in order to combat the lack of a support the gear arbor should be made as short as possible so as to place the gear blank as close to the dividing head front bearing as possible. Fig. 62 shows such an arbor.

				Angle of Inclination				
DP of Gear	30°	35°	40°	45°	50°	55°	60°	65°
				Cross Slide in-Feed				
6	.416	.439	.470	.509	.560	.628	.720	.852
8	.312	.330	.352	.382	.420	.471	.540	.639
10	.249	.264	.282	.305	.336	.377	.432	.511
12	.208	.220	.235	.256	.280	.314	.360	.426
14	.178	.188	.201	.218	.240	.268	.308	.364
16	.156	.165	.176	.191	.210	.235	.270	.319
18	.139	.146	.157	.170	.187	.209	.240	.284
20	.125	.132	.141	.153	.168	.188	.216	.256
22	.113	.120	.128	.139	.152	.171	.196	.232
24	.104	.110	.117	.127	.140	.157	.180	.213
26	.096	.101	.108	.117	.129	.145	.166	.196
28	.089	.094	.101	.109	.120	.134	.154	.182
30	.083	.088	.094	.102	.112	.126	.144	.170
32	.077	.082	.087	.095	.104	.117	.134	.159

CUTTING SPUR GEARS IN THE VERTICAL MILLING MACHINE

In recent years more and more home workshops are being equipped with a vertical milling machine. The model engineer always expects his machine tools to perform tasks other than their purely nominal duties. Basically a lathe is a turning machine but with a little ingenuity and some accessories the lathe is capable of being adapted into performing operations far removed from plain turning. Not so far behind in versatility is the vertical milling machine, so it is not surprising that as its potential is discovered this type of machine tool is increasingly being introduced into the home workshop. Some manufacturers of small vertical milling machines, recognizing the needs of the amateur, are producing machines with a geared or slow speed drive, and a machine of this type is ideally suited for use as a gear cutter.

Fig. 63 shows diagrammatically how a vertical milling machine is set up to produce spur gears. The cutter is mounted onto a short stub arbor which fits into the main spindle of the machine and is held in place by means of a drawbolt that passes through the hollow spindle. The cutter is therefore mounted in close proximity of the main spindle bottom bearing and so is well supported. If the machine has a geared drive then the full speed range right down to 30 or 40rpm is readily available to the cutter. The downfeed movement of the milling machine quill will be fitted with a fine feed mechanism enabling the cutter to be positioned at any height required and then locked into that position thus preventing further movement.

The large, flat T-slotted table that most milling machines possess provides an ideal surface for mounting the dividing head. The majority of dividing heads are designed around a headstock providing both a good center height and also a large mounting surface, so no problems should arise in securing the dividing head of this type to the table. A worm and wheel dividing head is an expensive piece of equipment and maximum versatility will be a consideration when deciding which one to either purchase or construct. A dividing head is also a very useful lathe accessory and has many uses other than indexing gear blanks; however, when intended for use on a lathe it is usually designed for mounting onto a vertical slide and so does not possess either a large, flat base for direct mounting onto a machine table or a serviceable center height. The Myford head is a typical example of this type and although it is perfectly satisfactory for use on a milling machine it does need a raising block inserted between the head and the milling machine table.

In cutting spur gears it is essential to align the dividing head spindle parallel to the table slide-ways, otherwise the resulting gear teeth will not be square to the face of the gear. The photograph Fig. 64 shows the milling machine set up for cutting a spur gear; the milling machine shown has a locating slot or keyway in the table (this keyway was cut using the milling machine itself so as to guarantee the keyway being parallel to the slide-ways of the table). The raising block is also keyed on both top and bottom surfaces so that whenever the dividing head and raising block are fastened to the machine table they are automatically aligned correctly. It may be noticed that only

Fig. 64. *Cutting a spur gear on the vertical miller. The dividing head is mounted on a raising block*

one fixing bolt is used to secure the dividing head, this is perfectly satisfactory as both the raising block and head are located by keys and all the bolt has to do is provide a closing pressure.

If no automatic aligning system is incorporated in the clamping system then the dividing head will have to be set parallel each time it is fastened onto the table. The dividing head base may be provided with a datum face, in which case it is a simple matter of using a square between the datum edge and the table edge. If no setting datum is available then other methods will have to be used to obtain correct alignment. One method is to place the longest available arbor in the dividing head spindle and then, by operating the feedscrew handwheel, bring the cutter arbor into a position close to, but not touching, the long arbor and then lock the cutter arbor in position by means of the quill clamping device. Now, by moving the table to and fro and watching the gap between the two arbors widen and narrow, it will be obvious which way to move the dividing head to obtain the correct setting. Fig. 65 shows this operation being done.

The gear blank should be mounted onto a suitable arbor and this arbor also needs to be secured to the dividing head by means of a drawbar passing through the hollow spindle. Providing that the blank is positioned close to the dividing head main bearing and that there is little overhang then the arbor will not need a tail support. If, owing to some reason of design, the blank has a considerable overhang then it may be better to provide a tail support.

A commercially produced dividing head of the type for direct mounting onto a machine's table is usually supplied with a tailstock that has been specially matched to the head. This is to ensure that the center heights of both units are identical and that any workpiece placed between the two will be parallel to

Fig. 65. *Aligning a direct dividing head parallel to the table*

Fig. 66. *Supporting the cutter arbor by an independent but matching tailstock*

the table. The tail support is completely independent of the head and can be positioned at any convenient distance from the head. The photograph Fig. 66 shows a home-made direct division head with the cutter arbor supported by a tailstock. If a Myford-type of dividing head is being used then the support will be supplied by the overarm center. This is not as rigid as the table-mounted support but it does provide some assistance. However, if this proves to be insufficient then the tail of the overarm may be clamped to the table providing that a suitable packing is placed between the table and the support—refer to Fig. 67.

To centralize the cutter with the blank, withdraw the blank arbor from the dividing head and insert a lathe center in its place, then lower the head of the milling machine, or raise the table—depending on the type of milling machine being used—to within about a half-inch or so of the correct setting position, then lock the head firmly in position. Centralize the cutter by means of the fine adjustment downfeed mechanism. The whole process is similar to that described when dealing with centralizing in the lathe. Once the correct position has been obtained lock the quill very firmly, giving an extra tweak over the normal clamping as any movement of the quill during the cutting operation would certainly spoil the gear. By bringing the whole head down to the approximate setting position very little quill extension will be needed resulting in maximum support for the cutter—Fig. 68 is a photograph showing the centralizing being carried out.

Arranging for the cutter to cut towards the dividing head rather than away from it means that the cutter will be operating on the back side of the blank, i.e. the side away from the operator, and between the blank and the milling machine column. This is not a disadvantage as it will still be in view and if the machine has a variable throat depth similar to the machine shown in the photograph this depth will be set to the minimum overhang, allowing the cutting to take place as close to the column as possible. The depth of cut is applied by means of the cross-slide screw and, once again, it is best to cut the full depth of the tooth at one pass. The feed is applied by the table handwheel and if the machine is fitted with a power feed this may be used.

Fig. 67. *Clamping the tail of the overarm to improve rigidity*

Fig. 68. *Centralizing the cutter on a vertical milling machine*

CUTTING SPUR GEARS IN A HORIZONTAL MILLING MACHINE

All machine tools are designed primarily to perform some specific function and when the machine can be adapted to include other functions then the machine's versatility is increased. It is well-known that the lathe designed for model engineering heads the versatility list and so provides the backbone of the amateur workshop. A machine's general usefulness is reflected in its popularity and that is why the basic horizontal milling machine, or one without a vertical attachment, does not usually appear high on the amateur's list.

A horizontal miller is basically a production machine and is normally equipped with a long table traverse but usually it has a limited cross-travel. It is ideal for workpieces that require machining along one axis only but in the main it has far less potential than the vertical milling machine. Gear cutting is one operation, however, that is suited to the horizontal miller if it has the inherent stability and a suitable speed range to the spindle.

As the table of the horizontal miller is similar in size and shape to the table of a vertical machine, the arrangements of dividing head, work arbor and tailstock are fitted to it in precisely the same way. The main difference between the two millers when used for gear cutting is the attitude of the cutter. In a vertical machine the cutter revolves in the horizontal plane while in the case of the horizontal miller the cutter rotates in the vertical plane. This means that on the horizontal machine the cutter is set to machine the tooth on the top vertical center line of the blank and so the short stub arbor that was used in the vertical machine will not provide sufficient overhang to allow this position to be reached. In order to reach the center line of the dividing head and also allow a working clearance between the edge of the table and the head of the machine, a reasonably long cutter arbor will be required. This will necessitate the use of a tail support

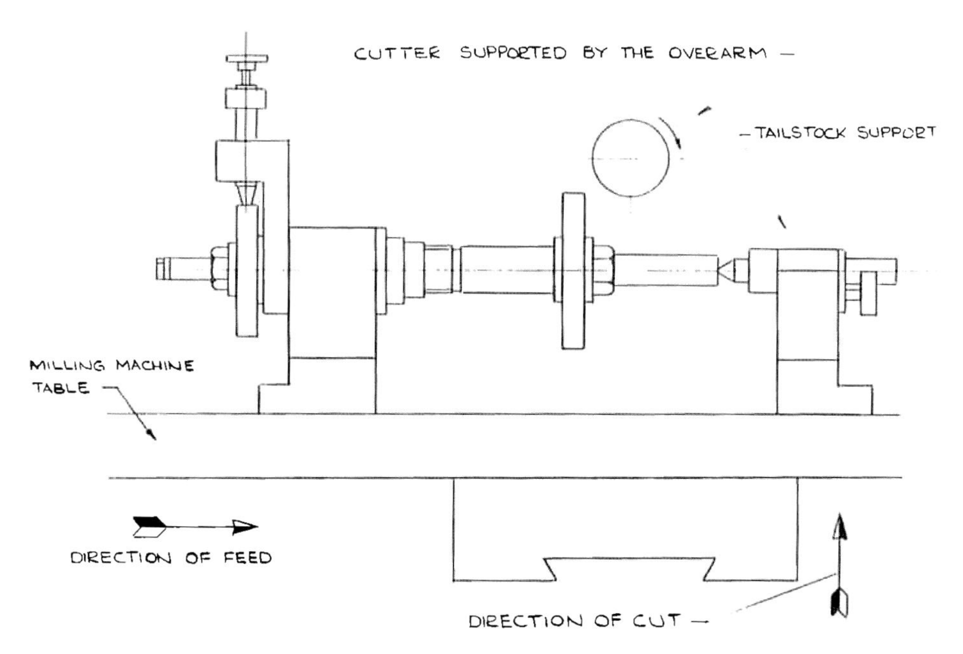

Fig. 69. *Cutting spur gears in the horizontal milling machine*

for the arbor, which can be provided by the overarm of the machine.

On horizontal millers the head is usually fixed so the cutter will have to be centralized by using the cross-slide feedscrew and the cut applied by raising the table. Both of these movements are then firmly locked and the feed applied by means of the table handwheel or table power feed unit. Once again, arrange the complete set-up so that the cutter rotates and is cutting towards the dividing head and not towards the tailstock support, Fig. 69 shows the arrangement diagrammatically.

Generally speaking, gear cutting in a milling machine, either vertical or horizontal, is preferable to cutting in the lathe—after all gear cutting is a milling operation and the milling machine is designed to withstand the type of forces involved. Also, with the milling machine the path between the cutter and the workpiece is much shorter and requires less components, which should be reflected in greater rigidity. The emphasis on rigidity is important as the length of cut involved is the distance around the profile of two adjacent teeth and this can be considerable, particularly when compared with the length of cut experienced in general machine work. So no matter which method or which machine is being used, always try to obtain maximum support for both the cutter and the workpiece and always cut towards the fixture that can best resist the thrusts developed.

It is also of great importance that both the cutter and the blank are correctly mounted, so therefore it pays to spend some time in both designing and constructing the arbors. Once made they can be used many times and any differences in the bores of blank and cutters can to some extent be accommodated by providing a bush or sleeve to increase the arbor's mounting diameter. Such a bush should be machined all over at the one setting and the hole produced by single-point boring as it is imperative to have both cutter and workpiece mounted without any eccentricity. As a matter of interest, the arbors shown in the accompanying photographs have been drawn out and are shown in Fig. 70.

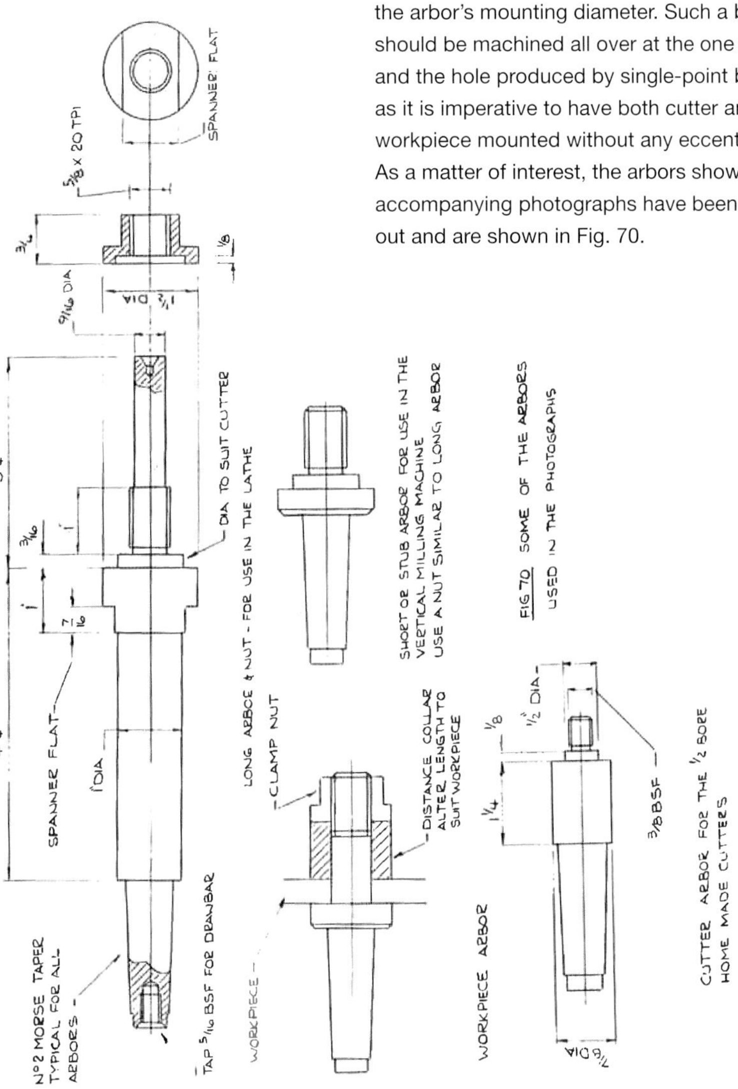

Fig. 70. *Some of the arbors used in the photographs*

CHAPTER 10
CUTTING WORMS AND WORM WHEELS

As stated in Chapter 6, a worm is basically a screw thread and so the machining of a worm is similar in principle to cutting a screw thread. This is done in the lathe with the cutter held in the toolrest in the normal way and the lead or helix angle of the worm is then generated by means of the lead screw driven by a train of suitable changewheels.

The tool used is naturally a form tool and, as previously stated, the tooth shape of the worm is basically similar to the rack form of the intended mating gear. If the mating gear is of involute form—as it most likely will be—then the cutter will be of a "V" shape with the included angle of the "V" being twice the pressure angle of the tooth profile used. As only two pressure angles are in general use the included angle will be 40° in the case of the popular 20° pressure angle, or 29° for the 14½° angle.

It is essential to grind the form tool to the correct profile and this is best done on a tool and cutter grinder. Tool and cutter grinders are becoming more popular in the amateur workshop as quite a number of designs are now available in kit form, enabling the amateur, at modest cost, to make a machine tool that may otherwise be outside his workshop budget. Not only are the "V" sides of the cutter important, the lip of the tool is also. Obviously it will not come to a point like a normal screw cutting tool but will have a flat end, and the width of this flat is important.

If a tool and cutter grinder is not available then the tool can be ground to a suitable gauge by the "free hand" method but great care will be necessary in order to obtain both the correct profile and clearance angles. Fig. 71 illustrates a suitable tool. It is recommended that the tool should be produced from round section, high speed steel; a ¼" dia. will be satisfactory for all but the larger DPs when it may be necessary to increase the tool diameter to ⅜". The width of the tip is important; if it is too wide then not enough clearance will be provided between the top of the gear tooth and the bottom of the worm and the tooth would rub on the bottom of the worm, or could even prevent assembly of the components. If the tip is too narrow then this in itself will not be a great detriment, as it will only produce more clearance between the gear tooth and the bottom of the worm.

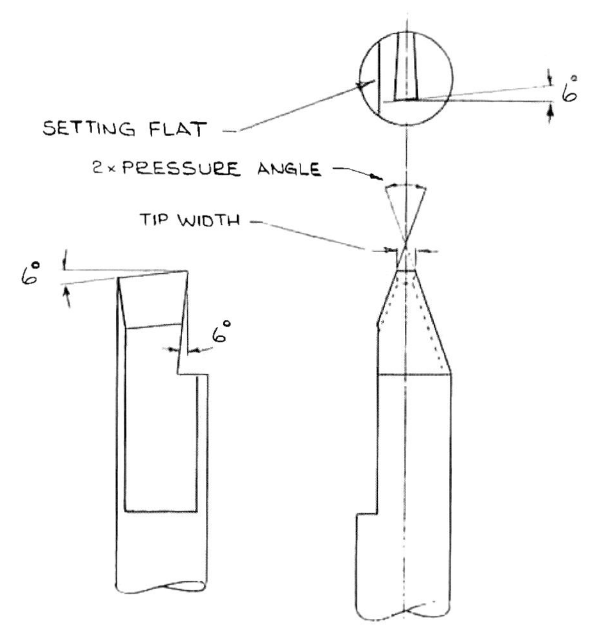

Fig. 71. *Form tool for cutting a worm—right hand shown*

However, should the tip be narrow then the depth of cut or infeed will have to be increased in order to produce the correct PCD on the worm. This can often be used to advantage with the one tool used to cut more than one DP, for instance, a tool made for, say, 20DP could also be used for 18DP or even 16DP provided that the correct infeed is used. In most cases the extra clearance provided will be of no consequence.

It is not easy to produce the correct tip width as there is nothing tangible to measure. Well-equipped toolrooms would be provided with special equipment such as a "Shadowgraph" that would enable the tip to be compared to some known standard. The amateur will have nothing like this at his disposal and will have to do his best with the normal measuring equipment available. It is possible to measure the width of the tip with a micrometer and, although the degree of accuracy may be open to question, nevertheless it will be within the limits required for most general needs. The micrometer used must be in good condition, or at least it must have anvils with good sharp corners. If the corners of the anvils are at all rounded it will not be possible to obtain a reasonable reading.

In order to measure the tip width, first close the anvils of the micrometer until the tip of the tool will just, and only just, enter the gap, then apply a gentle force so as to push the tool into the gap. The lip of the tool will be lightly held and a resistance to rotary motion in either direction will just be felt. Close the gap between the two micrometer anvils by .001 and try again. If the first setting was the correct reading then the tip of the tool will not enter the gap and the slight resistance to

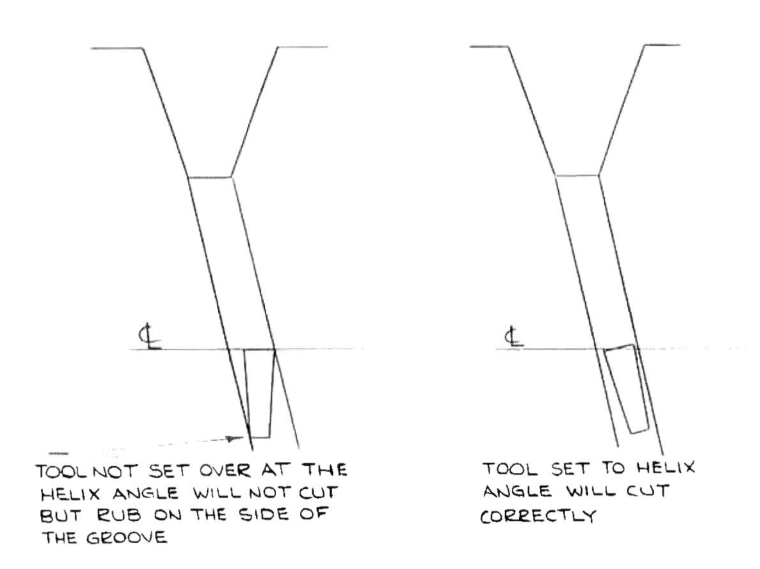

Fig. 72. *Showing the effect of the helix angle on the tool*

rotary motion will have gone. The process may sound somewhat hazardous but with a little practice a delicate sense of feel can be quickly acquired and surprisingly good gauging results achieved.

The worm, being a screw, will naturally have a helix angle and this angle will in most cases be far greater than those normally encountered in general screw cutting operations. In order for the tool to cut correctly it must be set over in its holder at the helix angle, or at a close approximation to the angle. Failure to do this will prevent the tool from cutting as the clearance and rake angles will be adversely affected and the side of the tool will rub on the side of the groove, see Fig. 72.

The reason for recommending that the tool be made from round material is now apparent—it can be ground in the normal way and then rotated in its holder until the required helix angle has been obtained. By referring to Fig. 71, it will be seen that a setting flat is recommended on the side of the tool. If this flat is produced normal to the tool vertical axis it can be used as a datum face and so simplify the setting process. A protractor set to the required angle and placed upon the flat top of the cross-slide will indicate how far to rotate the cutter in order to obtain the correct setting.

The following tables, listing a wide range of both DP and modular pitches, show the tip width for both 14½° and 20° pressure angles, the depth of cut to be applied and also how to arrive at the outside diameter of the worm blank given the PCD. The PCD will have been determined by the design requirements. The last column shows the linear pitch, which will be needed in order to calculate the correct train of changewheels to apply to the banjo and so drive the lathe lead screw.

DP	Tooth Tip Width		Depth of Cut	OD = PCD +	Linear Pitch
	14½°	20°			
6	.162	.121	.360	.333	.5236
8	.122	.091	.270	.250	.3927
10	.097	.073	.216	.200	.3142
12	.081	.061	.180	.167	.2618
14	.069	.052	.154	.143	.2244
16	.061	.046	.135	.125	.1963
18	.054	.040	.120	.111	.1745
20	.049	.036	.108	.100	.1571
22	.044	.033	.098	.091	.1428
24	.041	.030	.090	.083	.1309
26	.037	.028	.083	.077	.0208
28	.035	.026	.077	.071	.1122
30	.032	.024	.072	.067	.1047
32	.030	.023	.067	.062	.0982
36	.027	.020	.060	.056	.0873
40	.024	.018	.054	.050	.0785

It is not possible to list the helix angles in these charts as this will depend upon the diameter of the worm. It has already been said that the diameter of the worm has no effect on the gear ratio and so the PCD of the worm can be chosen to suit the design requirements. However, it is not difficult to determine the helix angle as it is only a case of dividing the lead by the circumferential length of the pitch circle and then reading the angle from a set of tangent tables. The mathematical method is shown in Fig. 73.

Module	Tool Tip Width				Depth of Cut		OD = PCD +		Linear Pitch	
	14½° PA		20° PA							
	Inches	mm	Inches	mm	Inches	mm	Inches	mm	Inches	mm
4	.153	3.89	.115	2.91	.340	8.63	.315	8.00	.4947	12.57
3	.115	2.92	.086	2.19	.255	6.47	.236	6.00	.3711	9.42
2.5	.096	2.43	.072	1.82	.212	5.39	.197	5.00	.3092	7.85
2	.077	1.94	.057	1.46	.170	4.31	.157	4.00	.2474	6.28
1.5	.057	1.46	.043	1.09	.127	3.23	.118	3.00	.1855	4.71
1.25	.048	1.22	.036	.91	.106	2.70	.098	2.50	.1546	3.93
1	.038	.97	.029	.73	.085	2.16	.079	2.00	.1237	3.14
0.8	.031	.78	.023	.58	.068	1.73	.063	1.60	.0989	2.51
0.7	.027	.68	.020	.51	.059	1.51	.055	1.40	.0866	2.20
0.5	.019	.49	.014	.36	.042	1.08	.039	1.00	.0618	1.57
0.4	.015	.39	.011	.29	.034	.86	.031	.80	.0495	1.26
0.3	.011	.29	.009	.21	.025	.65	.024	.60	.0371	.94

The form tool must be held in a suitable tool holder, the basic shape of which may be square or rectangular, its actual shape depending upon the type of tool box with which the lathe is equipped. The holder must have a hole, preferably reamed to fit the tool bit, and also a means of firmly securing the tool in the holder after it has been set to the helix angle. Securing the cutter by means of a clamp screw passing through the side of the holder and making contact with the cutter is not recommended as this does not provide a firm or secure hold. The screw would merely push the tool over to one side of the hole, thus providing a clearance on the screw side and so allowing the tool to rock under load. It is far better to split the holder across the center line and then position the clamp screw so that when it is tightened the hole will close completely around the tool, thus making the whole unit rigid. Another advantage with this method is that there are no rotating forces placed upon the tool during the clamping process so once the tool is set to the helix angle it will not rotate as the clamp screw is tightened.

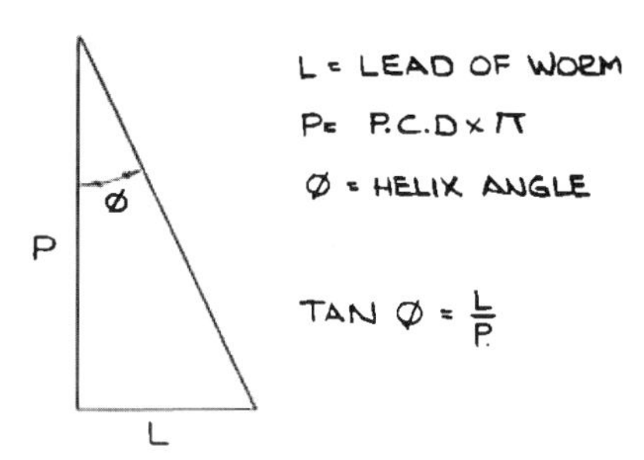

Fig. 73. *How to calculate the helix angle*

DETERMINING THE CHANGEWHEELS TO PRODUCE THE HELIX ANGLE

When screw cutting in the lathe the helix angle of the thread is generated by gearing the lead screw to the lathe mandrel. The ratio of the gears that drive the lead screw can be adjusted by altering the changewheels on the lathe banjo so that the amount the tool advances for each revolution of the workpiece can be predetermined. Most lathes are equipped with a screw cutting chart listing the changewheels needed for most standard pitches and, knowing what pitch is required, it is then a simple matter of looking at the chart and fitting the gears accordingly. Selecting the gears required to cut a worm may not be so simple as it is most unlikely that the pitch required will be shown on the list. Working through an example is possibly the best way of illustrating and solving the problems involved.

Supposing we have a 40-tooth spur gear and that we wish to cut a worm in order to construct a dividing head or rotary table. Fig. 51 shows how this can be done by setting over the worm to suit its helix angle. The first stage is to identify the gear positively and the only information that can be obtained by examination is the number of teeth and also its outside diameter. However, from this we can determine the circular pitch which will then give us the correct lead of the worm. It is most probable that the gear will be a standard DP gear and we know from previous discussion that the DP of a gear is the number of teeth plus two, divided by the outside diameter of the gear. If, on measuring, we find that the outside diameter is 1.400 then by applying the equation we get: $DP = \frac{40 + 2}{1.4} = \frac{42}{1.4}$

which gives us the figure 30 for the DP. However, in order to cut the worm we require the circular pitch so, again referring to previous discussion, we know that the CP is π/DP so that the CP for our gear will be $\pi/30$ which is .1047. We now know that we have to cut a worm with a pitch of .1047. Imperial screw cutting or changewheel charts rarely give the pitch but quote the threads-per-inch. If we wish we can change the pitch to threads-per-inch by dividing the .1047 into 1 to give 9.551 threads-per-inch. It is useless looking at the lathe chart for a train of changewheels to cut this number as it will not be listed! We are, therefore, on our own and must start with first principles. It will be better if, before trying to solve this non-standard pitch problem we examine how to deal with a standard pitch.

One of the most popular pitches used in model or small work is 40 t.p.i. so we will work out a set of changewheels for this pitch. Most small lathes are fitted with an 8 t.p.i. lead screw and this will have a pitch of ⅛" or .125". This means that when the half-nuts are engaged the lathe saddle, and hence the tool, will move a distance of .125" for every revolution of the lead screw. A 40 t.p.i. screw has a pitch of 1/40" or .025" and so the lead screw will only have to rotate a part of a turn in order to produce this movement. It follows from this that the lead screw will have to rotate at a slower speed than the mandrel and that the ratio between these speeds must be the same value as the ratio between the two pitches. This ratio is $.125/.025$ or 5:1, so this must be the speed ratio between the two shafts. The lathe mandrel will have to rotate five times for every single turn of the lead screw. In operation this will produce five threads on the workpiece over a distance of .125" so the pitch of each thread is .025".

We know now that the ratio of the gear to be placed on the banjo must be 5:1. The smallest gear in a set of changewheels is usually the 20-tooth gear so multiplying both top and bottom of the ratio by 20 gives us $\frac{5}{1} \times \frac{20}{20} = \frac{100}{200}$. If these two gears, together with an idler to bridge the gap, could be used the 20-tooth would fit on the lathe mandrel and the 100-tooth on the lead screw and the desired pitch would be generated. The snag here is that a 100-tooth gear is not often included in a standard set of changewheels so an alternative may have to be found.

The ratio of 5:1 can be subdivided into two, one being 2½:1 and the other 2:1 these two together would result in an overall reduction of 5:1. There are quite a few ways of obtaining the 2:1 from a set of changewheels as follows: $(40)/(20)$ or $(50)/(25)$ or $(60)/(30)$ or $(70)/(35)$. The 2½:1 ratio can be obtained by the gear $(50)/(20)$. By using four different sized gears a suggested train could be $\frac{50}{20} \times \frac{60}{30}$, which if multiplied out gives $3000/600$ or $5/1$. The gears shown on the upper line would be the driven gears while the gears shown below would be the drivers.

It has been stated earlier that in a train of gears only the first and last gear have any effect on the overall ratio, and this is so, but only if all the gears are on separate shafts. In the ratio shown above all the gears would not be on separate shafts; two of them would be on one shaft or compounded, and the sketch Fig. 74 shows how this is arranged. The 20-tooth gear drives the 50-tooth gear, which is fastened onto a common shaft with the 30-tooth gear so they revolve as a unit and must therefore rotate at the same speed. In this way all four gears affect the overall gear ratio.

There are other ways in which the four gears can be arranged and still maintain the 5:1 ratio and so long as the 20- and 30-tooth gears are used as drivers with the 50- and 60-tooth gears as the driven, the overall ratio will not be affected. Fig. 75 shows three more ways of arranging the four gears. Sometimes, although it is theoretically possible to arrange the gears in a particular way there may be practical difficulties owing to a large gear interfering with the securing nut of a smaller gear. However, re-arranging the train usually overcomes this problem. Generally speaking it is advantageous for the smallest gear to be used as the first driver and the largest

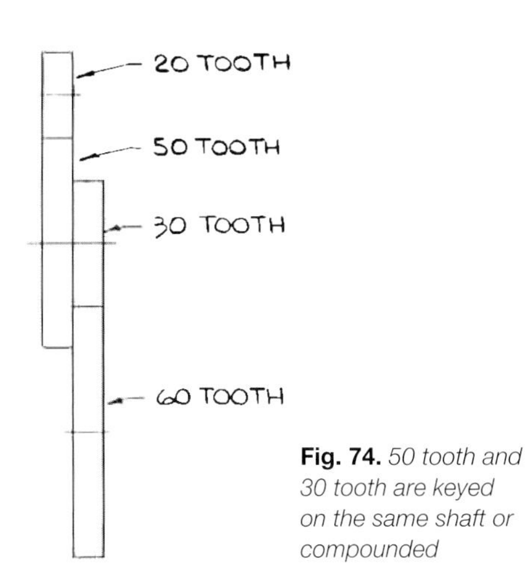

Fig. 74. *50 tooth and 30 tooth are keyed on the same shaft or compounded*

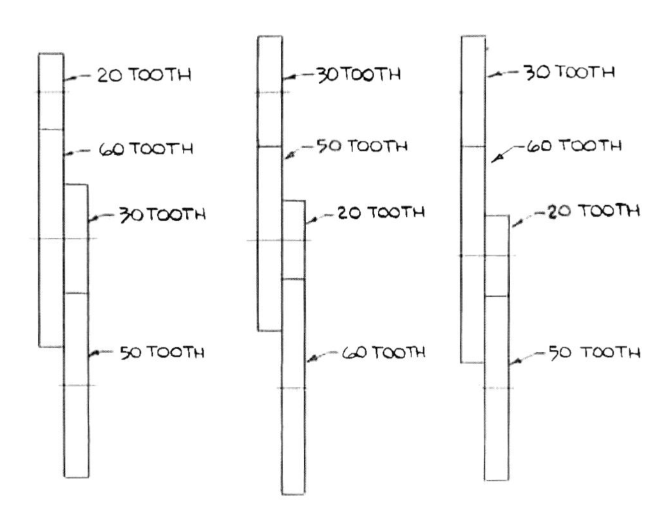

Fig. 75. *Showing three more ways of arranging the four gears shown in Fig. 74 without affecting the gear ratio*

gear to be the last driven gear but this is not mandatory.

Going back to our original problem of the .1047 pitch, it is dealt with in the same way as the standard pitch discussed. The first stage is to determine the gear ratio, and this is done by dividing the required pitch into the lead screw pitch. $\text{Ratio} = \frac{.125}{.1047} = 1.194 : 1,$ or $1/1.194$ But it is only possible to work in whole numbers when dealing with the number of teeth in a gear, so therefore the top and bottom lines must be multiplied by 1000:

$$\frac{1 \times 1000}{1.194 \times 1000} = \frac{1000}{1194}.$$

It follows from the previous example that two gears, one having 1000 teeth and one having 1194 teeth, will give us the pitch we require. Naturally two such gears are completely impractical but it is possible to simplify the fraction to $500/597$ and although the 500 looks promising, the 597 after reducing to 3 x 199 cannot be made any simpler, as 199 is a prime number and so to obtain the pitch required we must have a gear with 199 teeth. It would appear that it is not possible to cut a worm that will mate exactly with a 30DP gear and this is correct. However, the operative word in the above statement is "exactly."

Fortunately, although mathematics is an exact science, engineering practice is not and if it were so nothing could ever be made! An engineering drawing may call for a component to be 1" in diameter but it is not a practical proposition to produce it to 1" exactly, and even if it were it would only be exactly 1" at one specific temperature. There is, however, always a tolerance band within which the component is acceptable and the broader this band, the easier it will be to comply with the specified tolerance. The size of the tolerance band is usually a compromise between what is ideally required and what can be satisfactorily produced. The designer and drawing offices will decide on the compromise, fix the tolerance band and then the component will be produced accordingly.

The amateur cannot work this way as it is pointless deciding on a tolerance if the equipment available cannot produce the component within the limits fixed. What the amateur has to do is determine what can be made and then decide if this is acceptable. In our example we know we cannot set a train of wheels to give the exact pitch required so we have to find how close we can get to the theoretical pitch and then ask ourselves if this is good enough?

We know we cannot use the fractions $1000/1194$ or $500/597$, so some other fraction that can be of use must be found and the closer the value of the new fraction to the original the smaller will be the error introduced. There is a mathematical way of finding the new fraction, a process called continuous fractions. Since the author set out to produce a book for the practical enthusiast and to keep mathematics down to a minimum he is not going to describe here how or why the method works. Should any reader wish to study the subject in detail he must refer to the relevant mathematical textbook, but provided the method set out below is followed step-by-step with no questions asked as to why this or that is done, then it is a matter of simple arithmetic! Although we are using one set of numbers in the following exercise it is the method that is important and if this is followed then any other numbers can be substituted and a satisfactory result obtained.

The first stage in finding a suitable approximate fraction is to perform a series of divisions starting from our original fraction $1000/1194$. The slightly simpler $500/597$ could be used, and normally would be, but as the main idea here is to demonstrate the method we will use the $^{1000}/_{1194}$ fraction. As it happens this fraction could be canceled down but other sets of figures may not do so. However, it does not matter which fraction is used, the original or the reduced, the end result will be exactly the same.

Start by dividing the top line (1000 in this case) into the bottom line setting it out in a similar way to normal long division:

```
1000)   1194  ( 1
        1000
        ----
         194
```

The 1000 is divided into the 1194 giving an answer of 1 with a remainder of 194. Now the similarity with long division ends because the next stage is to divide the remainder into the previous divisor, so we divide the 194 into the 1000 and the example now looks like this:

```
1000)   1194  ( 1
        1000
        ----
         194  1000  (5
               970
              ----
                30
```

The 194 is divided into the 1000 giving an answer of 5 with a remainder of 30. The next stage is to repeat this process dividing the remainder, 30, into the previous divisor, 194, which will give an answer of 6 with a remainder of 14 and then keep on repeating this division process until the remainder becomes zero.

```
1000) 1194 (1
      1000
      ----
       194) 1000(5
             970
            ----
              30)  194(6
                   180
                  ----
                    14)30 (2
                           28
                          ---
                               2) 14 (7
                                  14
                                  --
                                   0
```

In our example we have arrived at the zero remainder with only five divisions, with other figures it may take more, seven divisions seeming to be an average number, but it may take as many as nine or so. The only information we require from the divisions is the series of successive quotients, viz:- 1, 5, 6, 2, 7.

If the reader now repeats the process but this time using the simplified $^{500}/_{597}$ fraction he will arrive at exactly the same series of quotients.

From this series of quotients we are going to produce a series of fractions and the first one is always the same regardless of how many quotients there are or what they may be, the first fraction must be $^{0}/_{1}$. This is followed in sequence by 1 divided by the first quotient in the series, which in our example is also 1. So our first two fractions are $^{0}/_{1}$ and $^{1}/_{1}$. Now for the tricky part! Multiply the numerator (top line) of the last fraction ($^{1}/_{1}$) by the next quotient in the series (5) and add the numerator of the previous fraction ($^{0}/_{1}$), which gives us 1 x 5 + 0 = 5. Do exactly the same with the denominators (bottom line), multiply the denominator of the last fraction ($^{1}/_{1}$) by the quotient just used (5) and add the denominator of the previous fraction ($^{0}/_{1}$), giving 1 x 5 + 1 = 6. This then gives us the next fraction in the series $^{5}/_{6}$. Repeating this procedure until all the quotients have been used we get:-

$$\frac{0}{1} : \frac{1}{1} : \frac{5}{6} : \frac{(5 \times 6 + 1)}{(6 \times 6 + 1)} = \frac{31}{37} : \frac{(31 \times 2 + 5)}{(37 \times 2 + 6)}$$

$$= \frac{67}{80} : \frac{(67 \times 7 + 31)}{(80 \times 7 + 37)} = \frac{500}{597}$$

Our series of fractions is therefore:

$$\frac{0}{1} \quad \frac{1}{1} \quad \frac{5}{6} \quad \frac{31}{37} \quad \frac{67}{80} \quad \frac{500}{597}$$

and it can be seen that the final fraction in the series is the same as the fraction that initiated the whole process, and this is as it should be, for no matter what the starting fraction may be, the final fraction in the completed series will always be of the same value as the original fraction. This is fortunate and gives us a check on our calculations as well as proving that all the intermediate fractions are also correct. If the final fraction is not the same as the original starting fraction, or a reduced version of it, then a mistake has been made in which case it is pointless to proceed further until the error has been rectified.

What we have in fact produced is a series of fractions starting with a rough approximation of the original and with each consecutive fraction becoming increasingly closer to the value of the original until, finally, we actually reach the original. It is surprising just how soon a close approximation is reached. Ignoring the first two fractions, which are only used to start the process, we get $\frac{5}{6}$, which is .8333; the original $\frac{1}{1.194}$ has a value of .8375 so the error is only just over .004. The next fraction in the series is $\frac{31}{37}$, showing an error of just over .0003, while in the case of $\frac{67}{80}$ the fifth place of decimals is reached before any error is found. It can be seen from this that if we can use one of the fractions towards the end of the series to determine the changewheels then the pitch error generated is going to be very small indeed.

Looking along the line the $\frac{5}{6}$ is attractive because the two gears $\frac{50}{60}$ or $\frac{25}{30}$ combinations are the obvious choice. If we were to use this combination to drive the lead screw the pitch cut would be $\frac{50}{60} \times .125$ (lead screw pitch), which again is .10416. This shows a pitch error over the theoretical .1047 of only half-a-thousandth or so, which may be considered adequate if no further improvement can be achieved. But, by using one of the other fractions we know this can be improved upon and if the two wheels 67 and 80 were available this would be ideal as the pitch cut would be .1046875, giving an error of only .0000125, which is less than the pitch error of the lead screw!

Unfortunately, changewheels are usually supplied with teeth in multiples of five and so wheels such as 31, 61 and 67 will not normally be available. In fact 31, 61 and 67 are all prime numbers so these fractions cannot be transposed into changewheels and it would therefore appear certain that the best we can obtain is the $\frac{50}{60}$ combination.

However, fractions have one useful property as far as our problem is concerned and that is, if we add together—or subtract from each other—the top and bottom lines of two fractions of similar value we then obtain a third fraction of the same value. For example, $\frac{9}{18}$ and $\frac{6}{12}$ are both equal to $\frac{1}{2}$ and if we add the top lines together and also the bottom lines together we get $\frac{15}{30}$, which equals to $\frac{1}{2}$. Subtracting the same numbers gives us $\frac{3}{6}$, which again is equal to $\frac{1}{2}$. From this it follows that if we treat two fractions of nearly the same value in a similar way the new fraction obtained will have a value close to the original two—in fact its value will lie between the values of the two fractions used.

Using this knowledge then we can supplement our series of fractions still further. Although 31 and 67 are prime numbers, if two prime numbers are added together the result must be a number that has factors. Therefore we can obtain a new fraction from $^{31}/_{37}$ and $^{67}/_{80}$, which gives us $^{98}/_{117}$. Now this looks more promising, as both top and bottom figures will factorize thus: $\frac{2 \times 7 \times 7}{3 \times 3 \times 13}$ or, using four numbers only, $\frac{14 \times 7}{9 \times 13}$; then by multiplying the 7 and 9 by 10, and the 14 and 13 by 5, we get the equation $\frac{70 \times 70}{90 \times 65}$. This would necessitate using two 70-tooth wheels and a 90-tooth wheel if available. Only one 70-tooth wheel would be available in a standard set of changewheels but both the 70 and the 90 can be divided by two giving a final gear train of $\frac{35 \times 70}{45 \times 65}$ and these are all standard changewheels. The pitch such a train would generate is $\frac{35 \times 70}{45 \times 65} \times .125 = .1047008$, giving an error of .0000008, which is much smaller than the error obtained by using the 67/ 80 combination. The reason for this is that the error in each successive fraction of the original series alternates between a plus and a minus value so that when the two are added together the error tends to be canceled out. It may not always be possible to arrive at a solution with an error as small as this using only standard changewheels but it is nevertheless surprising just what can be achieved by a little manipulation of the figures. Should a special changewheel be needed then it may be worthwhile producing one as it will increase the available range for future requirements.

This method of selecting changewheels for irregular pitches has been explained in depth because it can be used in solving any pitch problem where other methods have been found unsuitable.

In cutting worms to mesh with a DP gear the irregular function in the calculation is the factor π. A close approximation of π in fraction form is $^{22}/_{7}$ so if both top and bottom of this fraction are multiplied by 5 we arrive at $^{110}/_{35}$, which can be expressed in terms of changewheels as $\frac{55 \times 2}{35 \times 1}$. Going back to our original example of the 30DP worm, if this fraction is used in the calculation rather than a more exact decimal equivalent we can arrive at a set of changewheels directly without having to use the continuous fraction method, although the pitch error may be greater.

$CP = \pi$, $CP = \frac{55 \times 2}{35 \times DP}$, so for the 30DP we get $\frac{55 \times 2}{35 \times 30}$:

The required ratio =

$$\frac{\text{Leadscrew pitch} \times 35 \times 30}{55 \times 2} :1$$

or, $\frac{55 \times 2}{.125 \times 35 \times 30}$

Removing the decimal, we get

$$\frac{55000 \times 2}{125 \times 35 \times 30}$$

and simplifying this gives

$$\frac{440 \times 2}{35 \times 30} \text{ or } \frac{55 \times 16}{35 \times 30}$$

So the suggested train becomes $\frac{55 \times 40}{35 \times 75}$

and if we multiply this train by the 35 x 75 lead-screw pitch we get a figure of .1047619. Comparing this with the pitch generated by the train of wheels suggested by the continuous fraction method we see that the error, although greater, is nevertheless correct to the fourth place of decimals and for most applications this may be considered satisfactory.

The following table has been prepared suggesting the train of wheels that can be used to produce worms to suit the DPs listed. The trains have been calculated using the $\pi = \frac{22}{7}$ method and as can be seen all the trains incorporate the two gears 55 and 35, this being the π factor. All the gear trains are calculated for a lathe fitted with an 8-tpi lead screw. The top numbers are the drivers and the low numbers the driven wheels.

DP	GEAR TRAIN
6	$\frac{55 \times 80}{35 \times 30}$
8	$\frac{55 \times 60}{35 \times 30}$
10	$\frac{55 \times 40}{35 \times 25}$
12	$\frac{55 \times 40}{35 \times 30}$
14	$\frac{55 \times 40}{35 \times 35}$
16	$\frac{55 \times 40}{35 \times 40}$
18	$\frac{55 \times 40}{35 \times 45}$
20	$\frac{55 \times 40}{35 \times 50}$
22	$\frac{55 \times 40}{35 \times 55}$
24	$\frac{55 \times 40}{35 \times 60}$
26	$\frac{55 \times 40}{35 \times 65}$
28	$\frac{55 \times 40}{35 \times 70}$
30	$\frac{55 \times 40}{35 \times 75}$
32	$\frac{55 \times 30}{35 \times 60}$
36	$\frac{55 \times 20}{35 \times 45}$
40	$\frac{55 \times 20}{35 \times 50}$

In all the foregoing no account has been taken of the effect on the worm pitch of setting the worm over to the helix angle in order to obtain a mesh with the spur gear, Fig. 51. Theoretically, this set-over affects the pitch as an examination of Fig. 76 will show. The pitch we have been cutting is the distance marked "P," which is the circular pitch of the gear. To be correct the pitch of the worm is the distance marked "A" measured along the center line of the worm. In the illustration the helix angle has been exaggerated in order to illustrate the point.

In practice, if the helix angle is kept down to about 3° or so then for most applications a correction factor need not be applied as its effect will be very small. For instance, going back to our example, if the helix angle of the worm had been 3° the effect on the pitch of the worm would have been .00014, which is less than most model or amateur engineers care to worry about. Should the design requirement necessitate the introduction of a correction factor then it is simply a matter of dividing the circular pitch by the cosine of the helix angle of the worm to obtain the new pitch. The resulting gear train would then have to be calculated by the continuous fraction method. If the worm wheel is a helical gear and not a spur gear then the axis of the worm will be at right-angles to the axis of the gear, but the pitch correction factor still applies to the worm, the only difference being that in this case it will be the teeth of the worm wheel that will be inclined and not the axis of the worm.

In producing a worm the actual screw cutting procedure is basically similar to any other single-point screw cutting process. Some people prefer to use the cross-slide only

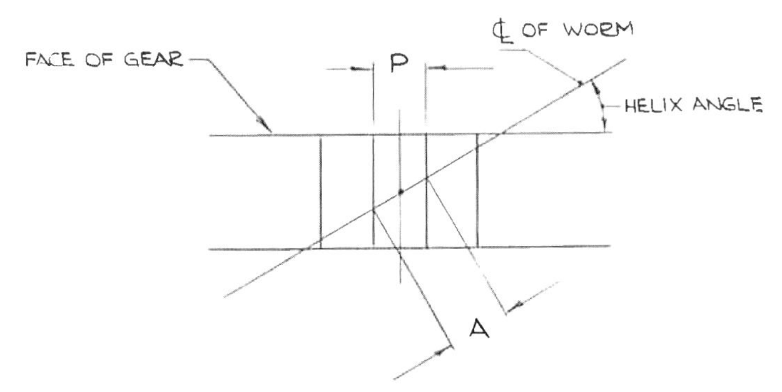

Fig. 76. *Showing the effect on the pitch of the worm by setting over at the helix angle*

for both applying the cut and withdrawing the tool, in which case the tool becomes a true form cutter and it will cut all the way around its profile. Another method is to set the top slide over to a half of the thread-angle and apply the cut by means of the top slide, using the cross-slide for withdrawing the tool at the end of each pass then, after winding the saddle back, returning the tool to its original position before applying the next cut. A combination of the two methods can be used, juggling with the two handwheels until the correct depth of cut has been achieved. The first method is quite satisfactory if the pitch of the screw is fine and the depth of the cut shallow but it is not recommended for cutting worms where it is unlikely that either of these conditions will apply.

The method favored by the author, and used with satisfaction when cutting deep, coarse pitch threads, is to set the top slide over but not to the "correct" half-thread form angle—just slightly finer. The cut is applied by the top-slide with the cross-slide used in the normal way for withdrawing and repositioning the tool. This method allows the leading edge of the tool, which has positive rake, to remove the bulk of the material but still leaves a fine shaving for the other side of the tool. This method was advocated and described in *The Model Engineers Workshop Manual* by Mr. G. H. Thomas.

The depth of the cut is obviously important as this will determine the size of the PCD. Using the suggested cutting method will make it difficult to obtain the depth by use of the handwheel micrometer dials so some other method may have to be employed. Perhaps the simplest and most positive way is to provide the worm blank with a short extension turned to the root diameter of the thread; it will then be obvious when the correct diameter has been reached.

It will not be possible to use the thread dial indicator to engage the lead screw for the start of each cut as, owing to the awkward pitch, a correct pick-up would not be shown on the dial. Once the half-nuts are engaged they must remain closed until the screwing operations have been completed. This will mean stopping the lathe at the end of each cut and rotating everything in the reverse direction until the starting position has been reached. The tool must be withdrawn first, however, as to try and reverse the lathe with the tool still in cut would result in disaster owing to that old enemy backlash.

The pitch of all DP based worms up to and including 24DP will be coarser than the pitch of the lathe lead screw (assuming an 8-tpi lead screw) and this means that to generate the desired pitch the lead screw will have to rotate faster than the lathe mandrel. This in turn will lead to high stresses being generated, particularly in the changewheels, so the speed of rotation must be set on the slowest possible speed. In fact it is a good idea not to use power at all but to rotate the mandrel by means of a handle fitted into its outer end, which will give greater control over the operation and also add a sense of feel. Myfords actually supply such a handle which will fit all their series 7 and 10 lathes. For the really coarse pitches it is better to actually fit a handle directly onto the end of the lead screw as this will lower the stress in the gears and make for easier turning. The best place to fix this handle is on the headstock end of the lead screw and not tailstock end as this will then eliminate a tendency for the lead screw to wind.

PRODUCING WORM WHEELS

Reference back to Chapter 6 will show that three types of worm wheels were described, the first of these being a simple spur gear suitable for slow-moving or low-power transmissions. The cutting of this type of wheel has already been described. The second type of wheel is basically similar to the spur wheel but the teeth are angled at the helix angle of the mating worm. From a power transmission point of view this wheel has little advantage over the spur gear although the angling of the teeth does allow for a simpler design of gear box, as the axes of both the components are at right-angles.

Producing this type of gear in the home workshop is not easy and it may not be possible without producing some additional tooling. It is not simply a matter of cutting the teeth at an angle because the teeth do not follow a straight line but are part of a helical curve. The wheel is in fact a thin slice of a long lead multi-start worm. It follows from this that to cut this type of worm wheel the helical form will have to be generated in a similar way to that described for cutting worms. The helix angle will be very large, being in fact 90° minus the helix angle of the mating worm, which will result in only a small radial movement of the workpiece as the cutting form tool passes over it. The lead screw will have to make a good many revolutions for each turn of the lathe mandrel and in order to obtain the required gear reduction at least six changewheels will have to be used although in many cases this may not prove sufficient. It is quite possible that some modification will have to be made to the changewheel banjo to allow a sufficiently large gear train to be mounted.

Alternatively, it may be better to incorporate a worm drive in the reduction gearing. Some form of indexing device will have to be incorporated in the arrangement but it is difficult to arrange this on the lathe mandrel in any of the ways normally employed when dividing in the lathe, as the mandrel is already being used to generate the helix. Possibly the simplest method of dividing is some form of direct division device on the arbor carrying the blank. This arbor could be a two-piece unit, the main part being a mandrel secured to the lathe spindle either on the screwed nose or in the Morse taper. Securely fastened to this would be some form of division plate which could be either a spur gear with the appropriate number of teeth or else a plate into which a circle of holes has been drilled. A sleeve free to rotate on this mandrel would carry the workpiece and also some form of detent to engage with the division plate. The cutting action will be basically similar to a shaping or planing machine rather than a turning operation and as only light cuts can be taken, and as it will not be possible to open the half nuts for the return stroke, the whole operation will be lengthy and prolonged.

A far quicker method of producing the teeth is to mill them using the standard Brown and Sharpe type of cutter. The method of producing the helix will be similar to the method just described but the cutter will have to be mounted and driven by an auxiliary cutting head mounted on the cross-slide. The teeth can then be cut to depth at one pass but it will be necessary to withdraw the cutter at the end of each pass and after winding back and indexing, reset the cutter to the correct depth before cutting the next tooth.

As was the case with the spur gear, the milling machine is superior to the lathe for cutting helical gears but again the difficulty is in generating the helix angle. This is done in industry by means of a special dividing head called a spiraling head. This is a special form of dividing head in which the worm spindle is extended so that a gear can be keyed onto it. This gear is then connected to the milling machine table screw by means of a train of changewheels so that as the feedscrew of the table is rotated to provide movement to the table, the worm of the dividing head also rotates, which in turn rotates the main spindle of the dividing head and also the workpiece. The result is the cutting of a spiral or helix.

The lead of this spiral is determined by the ratio of the changewheels used. The dividing head is also provided with a device that permits its normal indexing function to be used without interfering with the spiraling drive. It is uncommon to find this type of dividing head, or the milling machine that is capable of accepting it, in an amateur's workshop but should any reader possess one then it is almost certain he will know how to use it. A detailed description of the head and its uses would be lengthy and involved and so is outside the scope of this book.

When producing blanks for helical gears the formula PD = N/DP is not correct as the helical gear has a greater PD and outside diameter than its equivalent spur gear. To understand the reason for this refer to Fig. 77. The "normal" pitch is shown as CP, this being the distance between two teeth at right-angles to the teeth but, owing to the angularity of the teeth, the equivalent distance around the face of the wheel is the distance marked "L." To arrive at the PD for a helical gear, divide the PD for a similar spur gear by the cosine of the angle shown as "A" on the drawing.

This description of cutting helical gears has been condensed because not many amateurs have the necessary equipment, nor the desire to make the equipment to produce these accurately. However, if the problems outlined

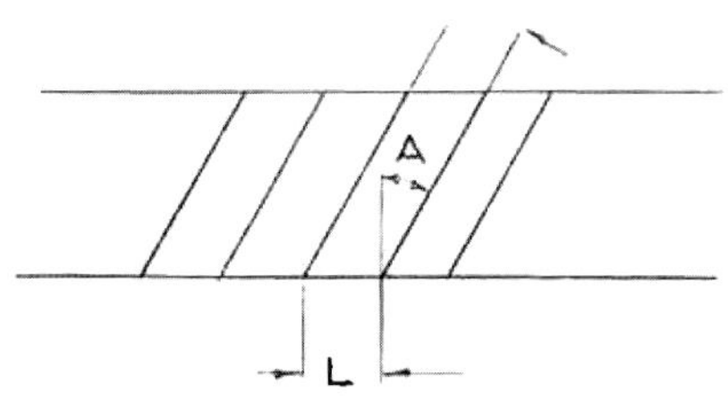

Fig. 77. *Showing why the PD and OD of a helical gear are larger than a corresponding spur gear*

have been understood then at least it will be a starting point for anyone wishing to pursue the matter further.

A worm wheel intended for power transmission, or one where it will be in operation for long periods of time, will have to have a long arc of contact similar to the example shown in Fig. 49. The worm wheel is hollowed out to fit over or wrap around the worm for a sixth or so of its diameter. This type of worm wheel cannot be cut using the normal type of gear cutter as the bottom of the teeth are not straight but curved to suit the worm and it has to be produced by a process known as hobbing.

The cutting tool, or hob as it is called, is similar in profile to the worm but has gashes cut into it to form cutting edges or teeth. If both this hob and the worm wheel blank are positioned on a fixture in their normal working attitude and both rotate at the same relative speed as if they were in mesh, then as the wheel is fed into the hob it will start to cut teeth into the wheel. The cutting action will be continuous because the wheel is rotating and so all the teeth will be cut at the same time. The hob is fed into the worm wheel until the required depth of tooth has been obtained, the cutting action is then complete.

Commercially, a special machine called a hobbing machine would be used and with this type of machine it is possible to rotate both cutters and the mandrel, holding the blank at any predetermined speed needed to suit the characteristics of the worm wheel being cut. Once again this is not a simple operation to carry out in the back-garden workshop but it can be done on the lathe providing a suitable attachment is made.

Rotating the cutter or hob is simple as this can be mounted either between centers or

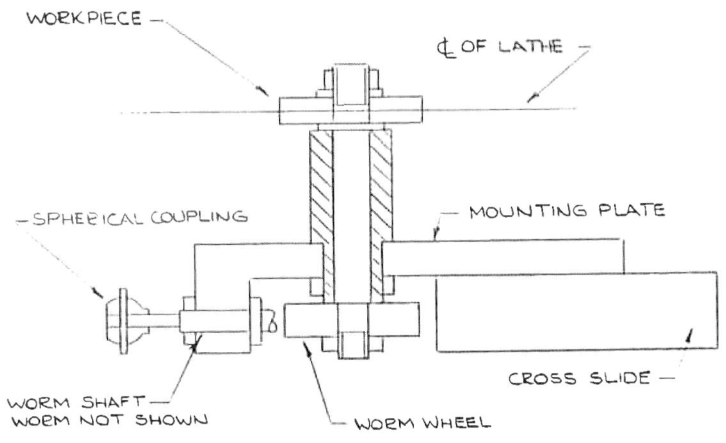

Fig. 78. *Fixture for hobbing worm wheel in the lathe*

held in a chuck and supported at the outer end by the tailstock center. The gear blank is mounted on a fixture secured to the lathe cross-slide, the main component of this fixture being a vertical shaft. The height of the shaft is such that when the blank is fastened onto it the center of the tooth face is coincident with the lathe center. The vertical shaft must be geared to the main lathe mandrel in such a way as to produce speeds relative to the speeds at which the worm and wheel will eventually operate. This means that if the ratio of the worm and wheel is 20:1, then the lathe mandrel must rotate twenty times to the gear blank's once.

To achieve this a worm wheel is secured to the vertical shaft, which is driven by a worm arranged in such a way that its driveshaft is parallel to the lathe bed and pointing towards the lathe headstock. Fig. 78 shows how this can be arranged. This particular worm shaft is driven from one of the changewheels on the banjo and the stud on which the changewheel is mounted must be hollow so that the spindle to which the changewheel is keyed can pass through it. The outer end of this shaft is provided with a simple coupling similar to the one shown in Fig. 78 and the drive shaft between the two couplings is simply a piece of tube whose inside diameter will pass over the spherical ends of the couplings.

Both ends of the drive shaft are slotted to provide a drive from the cross pins. If the drive shaft is made telescopic then it will be possible to adjust the position of the fixtures by moving the lathe saddle along the bed, thus allowing hobs of varying lengths to be used. The ratio between the worm and wheel on the fixture is not important so long as it is possible, with the changewheels available, to gear up the blank to the cutter at the ratio required. For instance, if the fixture

Fig. 79. *"Free" hobbing a worm in the lathe. The blank has first been "gashed" to remove most of the material and to correctly space the teeth*

has a worm wheel ratio of 20:1 and it is desired to cut a wheel with 30 teeth, then the changewheel ratio will be 2:3, so any wheel that will produce this ratio will be satisfactory, such as 20 and 30, or 30 and 45.

To anyone wishing to produce a worm and wheel it may seem rather strange to read that the first stage in their production is to acquire a pair! However, the worm and wheel used for the fixtures need not be any special size or ratio, in fact a worm and spur gear could be used. This would have a limited life but it could still hob many worm wheels before it needed replacing—it could even hob its own replacement which would be far superior to the original gear!

The hob should be made of carbon tool steel such as drill rod (or silver steel). This material is easy to obtain but, more importantly, it is simple to heat treat. It has already been mentioned that the hob is a facsimile of the worm, albeit suitably gashed to provide teeth, but like all cutting tools rake and clearance angles must be provided if the teeth are to be correctly cut. It is a simple matter to apply relief to a single-point tool such as a flycutter but in the case of a hob the relief will have to be produced by hand filing. This is quite tiresome and calls for considerable patience but unless it is done the hob will only rub and generate heat—naturally, this must be done before the hob is hardened. The hob will cut better if the gashes are in spiral form as this will always ensure that it will be in contact with the workpiece during the complete cutting operation. In industry hobs are form-relieved in a form-relieving lathe but it may be too much to suggest that the constructor should make such an attachment for his lathe, although it is quite feasible to do so.

FREE HOBBING

Mechanical rotation of the gear blank, as just described, is a highly desirable feature and the attachment is well worth constructing if any serious attempt at worm wheel cutting is contemplated. "Free wheel" hobbing, that is where the blank is not driven round under power but is free to rotate on its mounting spindle, is sometimes suggested. The idea is that as the cutting action is at an angle the blank will automatically revolve under the influence of the cutter. This is quite true and reasonable results can sometimes be obtained by this method, but it can be fraught with danger!

The author has carried out various tests using the free hobbing method and has produced some reasonable results but the main problem is that there is a "drag" on the workpiece, which tends to make the hob cut more on one side of the cutter than on the other, resulting in a continuous widening of the space between the teeth as the work proceeds. In some cases the result has been to lose a complete tooth and the wheel has finished with one tooth less than the required number!

For the method to work at all the hob must at all times be engaged with the wheel blank. This means that the teeth on the hob must be closely spaced, or preferably of a spiral tooth form. For this reason it is not recommended that an ordinary straight fluted tap be used as a hob because as soon as the trailing edge of the tooth leaves the blank, the blank will stop rotating and so the next cutting edge will not pick up the tooth and the process fails. Acceptable results can be obtained with spiral fluted taps but it is recommended that the final center distance between the worm and wheel is determined by direct measurement from the actual component after the hobbing has been completed.

If the worm tooth form is large enough then it is better to gash the blank first; this can be done with a slitting saw set at the approximate helix angle, the spacing being obtained by means of a dividing head. The hob does not then have to generate its own tooth space and, since much of the material has been removed, the "drag" is greatly reduced and the tendency to increase the tooth width is minimized. The wheel being cut in the photograph, Fig. 79, was gashed in this way.

CHAPTER 11

CUTTING BEVEL GEARS

"It is not possible to cut a correctly formed bevel gear with a form-cutter as the shape of a tooth is constantly changing along its length." That was a statement made in Chapter 5 and it is, of course, correct, as is also the remark made earlier that all engineering design is a compromise. Is it possible therefore to arrive at a compromise with regard to bevel gears and produce, if not a perfect, at least a serviceable profile using a form-cutter and a milling machine? The answer is "yes," providing it is realized that it will not be a precision gear. The amateur's needs seldom call for high precision, which is fortunate as it is usually a case of doing one's best with the limited amount of equipment available. However, such is the ingenuity of the model engineer that more often than not the final result is more than just satisfactory!

For many years bevels have been produced on milling machines. In fact before the development of generating methods, milling was the only machining process available and with the aid of a skilled fitter remarkable degrees of accuracy can be achieved. It is not possible to produce the tooth at one cut as its width tapers inwards towards the center of the gear, so the usual method used when cutting bevels on a milling machine is to base all the necessary calculations on the large end and try to produce the large end of the tooth to the correct theoretical profile.

The cutter used must be a special one, it cannot be the full width to suit the large end as it must be capable of passing through the small end of the tooth which is naturally narrower than the large end. The cutter, therefore, must be thin enough to pass through the small end without endangering its width. In order to obtain the correct tooth width at the large end the cutter must make two passes, one down each side of the tooth. The involute profile of the special cutter is correct for the large end so that after the second pass the tooth form at the large end is correct. Movement of both blank and cutter in between the two passes ensures that the width of the small end of the tooth is also correct, the snag being that the cutter has produced the same involute profile for the small end as it did for the large end. As the depth of the tooth is also less the net result is that the small end of the tooth will be "full," particularly on its face. This is where the fitter is employed. The teeth at the small end have to be dressed by hand filing in order to attain a serviceable shape and eliminate any tendency to bind, as would happen if the teeth were not dressed.

This method can be followed in the home workshop if the special cutters are available. A bevel gear of sorts can be made or "fudged" following the above method but using a standard cutter that is a compromise for either end of the tooth. After making various experiments the author is convinced that the best method to follow is to ignore the normal varying depth bevel and produce gears with parallel depth teeth.

PARALLEL DEPTH BEVELS

Parallel depth bevels are in fact a legitimate gear and were extensively used in the past, particularly during WW1, as they could be easily produced in jobbing shops on standard milling machines. With the advent of modern generating machine tools they no longer seem to be in favor but the fact that they can readily be produced on simple milling machines makes them an attractive proposition for the amateur.

As the name suggests, parallel depth bevels have a depth of tooth that is constant throughout its entire length but all other aspects of the tooth are similar to the standard gear form and meet at the apex point. All calculations are made from the small end of the tooth and not from the large end as was the case previously. A standard spurwheel cutter can be chosen to produce the small end shape at one pass and it is quite safe to traverse the cutter completely through the tooth because as the finished tooth becomes wider towards the outer edge, the cutter will leave a metal-on condition.

Further cuts, using the same cutter but with both blank and cutter repositioned, will remove this metal-on and open the tooth out to its correct width. This will produce a tooth with its correct curvature at the small end but as the tooth progresses towards the large end the curvature actually cut will be slightly smaller than the theoretical ideal. This will mean that the contact area between two teeth will become narrower at the large end but the teeth will roll together without any need for filing. In fact it is not a great detriment to have this condition and it is sometimes deliberately introduced into power transmissions to promote smoother running. The profile of the tooth at the large end will not be standard but of stub form and, again, this is a perfectly satisfactory profile and one used on certain special forms of gearing.

Bevel gears produced by this constant depth method do run together smoothly and require no additional dressing of any kind—such as running in with an abrasive to remove the "high spots."

In order to describe the actual method of cutting such a gear in detail the procedure adopted previously will be followed with a typical example worked through. The method is the same for all sizes and types so if the figures required are substituted for the figures quoted then a successful conclusion should be reached.

The example here is to produce a pair of miter gears of 20DP with 20 teeth and the first step is to accurately draw out the blank, as shown in Fig. 80. With miter gears we know that the pitch cone angle must be 45° from the center line and so this is drawn in first. The next stage is to determine the PCD and, as already stated, with this method all calculations are based on the small end.

Our teeth will therefore be 20DP at the smaller or inner face and the PCD,

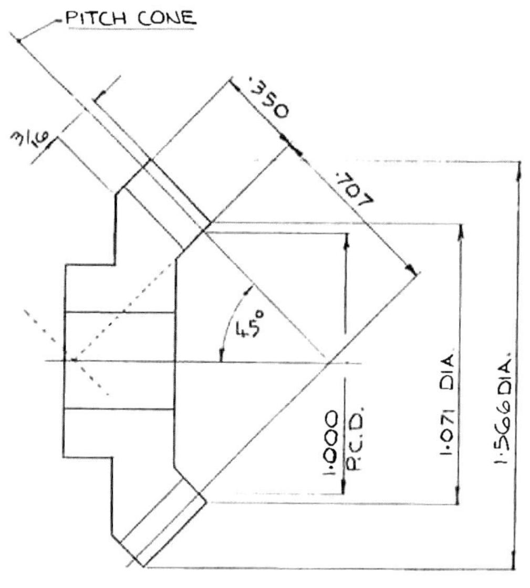

Fig. 80. *A 20-tooth, 20DP parallel depth miter gear*

CALCULATIONS.

PITCH DIA. - SMALL END :- $\frac{N}{DP} = \frac{20}{20} = 1.000$

O/DIA SMALL END $= PD + 2\ ADD^{N}$

$= 1.000 + (\frac{1}{DP} \times \sin 45 \times 2)$

$= 1.000 + (\frac{1}{20} \times .707 \times 2) = 1.071$

O/DIA LARGE END

$= [(.350 + .707) \times \sin 45 \times 2] + .071$

$= 1.566$

LARGE END DP $= \frac{N}{PD} = \frac{20}{1.566} = 12.77$

NUMBER OF TEETH ON BACK CONE PD. (FOR CUTTER SELECTION)

$= \frac{1}{\sin 45} \times 20 = 28.28$

USE 28 TOOTH CUTTER, WHICH IS No. 4

OFFSET FOR SECOND CUT

$= \frac{1}{2} \times$ CHORDAL THICKNESS $= \frac{1}{2} \times PD \times \sin \frac{90}{20}$

$= \frac{1}{2} \times 1 \times .078 = .039$

BLANK ROLL $= \frac{1}{4} \times$ INDEXING ANGLE

OR $\frac{1}{20} \times \frac{1}{4} = \frac{1}{80}$ REVS

which is N/DP, will be 20/20, or 1.000" diameter. This can now be marked on the drawing as can the line representing the inner face which will be perpendicular to the pitch cone. The distance from this face to the apex point can be found, as can any other profile dimension, by scaling the drawing. Reasonably accurate results can be obtained this way, particularly if the drawing is made four or five times full-size. However, it is quicker and more accurate to determine all sizes by direct calculation. In this instance it is half the PCD divided by the sine of the pitch cone angle, which gives us a length of .707". The length of the tooth face is simply a matter of good proportion and a face width of about one-third of the pitch cone length would suggest a figure of .350, which seems reasonable. This then fixes the back face of the teeth and also the large end PCD but as this is not required either for further calculations or in the machining processes, the actual figure need not be determined.

Next, draw on the addendum which we know from previous discussion is 1/DP, which will give us a figure of .050. It is now possible to determine the blank diameters for both

Fig. 81. *Ready to cut a bevel gear on a vertical milling machine. Note angle of dividing head*

large and small ends and the value of these, and the method used to calculate them, are shown on the drawing Fig. 80. The rest of the blank profile can be drawn to suit the individual requirements of the designer. The length of the back face is shown as 3⁄16" but there is no fixed method of determining this dimension as it has no bearing on the performance of the gear. It does however need to be greater than the tooth depth, which in this case is .108". This figure can be obtained from the tables as the depth of a 20DP tooth. No detailed instructions for machining the gear blank need be given here as this is purely a turning exercise.

The vertical milling machine is the best type of machine on which to cut bevels as it is far easier to set the dividing head at the required angle on this type of machine than on either the lathe or horizontal miller and so the description will feature the vertical milling machine. As with any other type of gear cutting it is important to select the correct cutter. Our bevel has 20 teeth and although the cutter to be used is a standard spur gear cutter it will not be the No. 6 cutter that would be used for a 20-tooth spur gear. The reason for this was illustrated when discussing bevel gear teeth in Chapter 5. The dotted lines in Fig. 80 show the bevel gear teeth "applied" to a standard spur gear and it can be seen from this that the equivalent spur gear is larger in diameter than the bevel and would therefore have to have more teeth. The cutter we require for the bevel is the cutter that would be chosen to cut this equivalent spur gear. The method of calculating this number is shown in Fig. 80 and comes out at just over 28; the correct cutter to use for this number is therefore the fourth cutter in the series which covers the range of 26–34 teeth.

The cutter is mounted on a short stub arbor and held in place in the miller spindle by means of a drawbolt, in exactly the same way described for spur gears. The same dividing head can also be used but instead of being set parallel to the table slide-ways it must be set over to the pitch cone angle, which in the case of this example is 45°; see the photograph Fig. 81 showing the setup ready for the cutting to commence.

Before the gear blank arbor is inserted into the dividing head, put a lathe center into it to centralize the cutter. Wind the cutter down to the central position by means of the downfeed handwheel; if any overshoot takes place wind the cutter back upwards sufficiently to remove any backlash in the system then wind down again. Do not approach the center position by "backing off," it must be approached on the downward movement. When the correct position has been reached lock the quill firmly in position and set the micrometer dial to zero.

The gear blank arbor can now be placed in position in the dividing head spindle and firmly secured by means of a drawbolt. Finally, the gear blank can be placed on the arbor and secured by means of a clamp nut. Manipulate the table and cross-slide handwheels until the cutter is just touching the blank then set the table micrometer to zero. Traverse the cross-slide inwards towards the back of the machine until the cutter clears the workpiece then by means of the table handwheel put on the cut to the full depth of .108" using the dial to obtain the correct setting, then lock the table very firmly in position.

Set the dividing head detent to the datum hole in any ring of holes that is divisible by four and cutting can then commence. The cutter will enter the blank by the small end and leave at the large end, which will give the machinist a good view of the cutting process. If a 60-tooth dividing head is being used, then in order to cut 20 teeth three turns of the handwheel will be needed to cut each tooth; carry on until all 20 teeth have been cut. This stage is shown in Fig. 82.

The tooth form will now be parallel throughout its length, being the shape required at the small end but still needing further cuts to widen the large end. Any additional movement to widen the large end of the tooth must not interfere with the small end profile; in other words, the position of the small end relative to the cutter must not be changed. The condition of each tooth will be as shown in the diagram, Fig. 84, and we will leave matters there while we consider a little theory.

Fig. 82. *The bevel wheel shown in the last photograph after the first pass of the cutter through all the teeth*

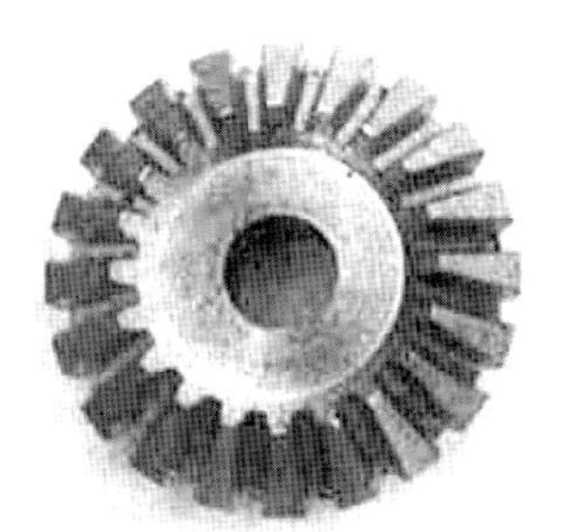

Fig. 83. *Two bevel gears: the one on the left after the first pass, the one on the right completed, having had three passes*

The lines representing the chordal thickness on the pitch circle are parallel to each other but in order for the gears to mesh correctly these lines must follow the dotted path shown in Fig. 84. They must be radial lines meeting at the apex point and passing through the chordal thickness points XX on the inner profile produced by the first, or gashing, cut.

Looking at the line AA, this is not a radial line, as it does not pass through the apex point, but if we were to move the cutter over to the right by a distance equal to one-half of the chordal thickness then the line would move with the cutter and occupy the position previously held by the center line and so would then pass through the apex point—it has in fact become a radial line.

Obviously, if a cut were to be taken through the blank after this cutter movement it would widen the tooth gap over its complete length and ruin the small end profile, but as the line AA is now a radial line, the whole blank can be rotated clockwise about its center until the two dots XX return to their original position relative to the cutter. The inner end of the tooth is now in such a position as will allow the cutter to pass through it without removing any more material and since the cutter is positioned on a true radial line it will produce one side of the tooth to the correct radial profile as it progresses through the cut. This second pass is now taken through all the teeth on the blank, after which each tooth will be the shape shown in Fig. 85.

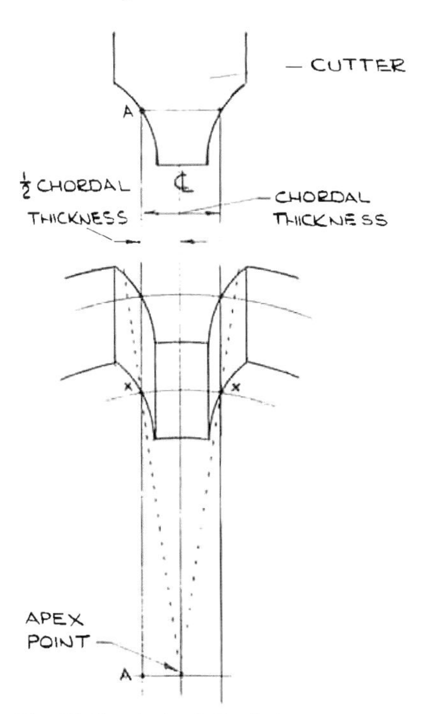

Fig. 84. *Bevel wheel after first pass*

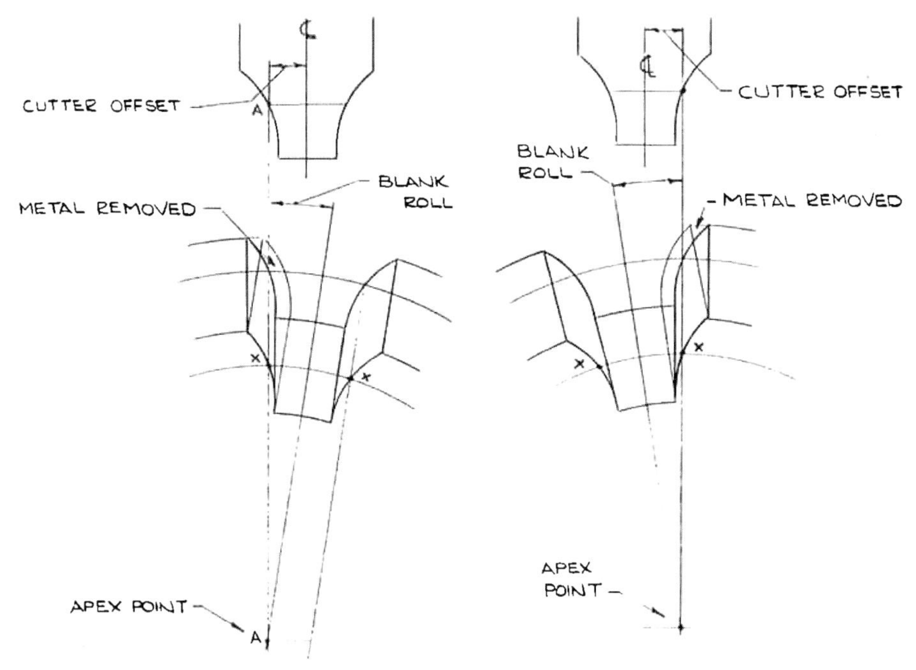

Fig. 85. *Bevel wheel after second pass*

Fig. 86. *Bevel wheel after third and final pass*

A third and final pass is needed to remove the surplus material from the other side of the tooth so that this also follows a true radial line. This is accomplished by first returning both cutter and blank to their original starting position and then by displacing them an equal amount in the opposite directions. The final tooth form will now be as shown in Fig. 86, with both sides of the teeth being truly radial.

Emphasis has been made about not removing any further material from the small end profile left by the first pass of the cutter, and the second and third cuts should pass through the tooth at this point without removing any further material. Theoretically this is impossible because the cutter in the second and third pass is at a slight angle to the original cut, which will result in a small shaving being removed from the top part of the involute curve. However, the amount of material actually removed will be barely detectable and have no significant effect on the performance of the finished gear.

Where a vertical milling machine is being used the points just discussed can be applied in a practical way. As the cutter on this type of machine rotates in a horizontal attitude the displacement of the cutter will be up and down rather than left and right. (Referring back to earlier text, this is the reason why it is important when centralizing the cutter to zero the micrometer dial with all the backlash eliminated.) So the first of the two movements needed to obtain the correct setting for the second cut for the 20-tooth bevel is to move the cutter down a distance of .039", using the fine feed mechanism of the quill, this dimension being obtained from the calculations shown in Fig. 80. The calculation is always the same and consists simply of multiplying half the pitch diameter by the sine of the angle, $\frac{90}{\text{number of teeth}}$. It is important to remember to lock the quill securely after resetting as any accidental movement during cutting would be disastrous.

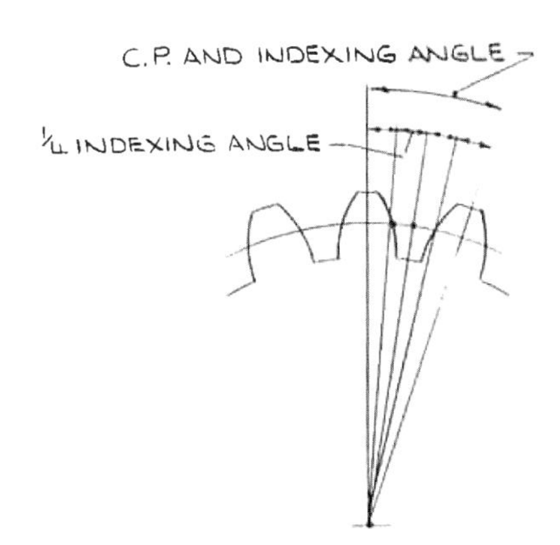

Fig. 87. *Showing why the blank roll is ¼ of indexing angle*

The second, or radial, movement is obtained by rotating the dividing head, and the amount of rotation required is equal to half of the tooth thickness on the pitch line. As a linear distance this would be an awkward measurement but as an angular displacement it is simply one-quarter of the tooth indexing angle—see Fig. 87. The gear being used as an example has 20 teeth, therefore the rotation of the dividing head will be one-quarter of this, or one-eightieth of a turn. On a 60:1 ratio dividing head this represents three-quarters of one turn, hence the need for a division plate capable of being divided by four. Once again, do not forget the eternal enemy of machinists—backlash, and when setting the dividing head back by three-quarters of a turn, go back at least one complete turn so that the desired hole is approached in the same direction of rotation as used in the initial indexing.

In order to obtain the correct settings for the third and final cut, return both cutter and dividing head back to their original or starting positions and then apply both off-set and blank roll but this time in opposite directions. The quill movement will be in the upwards direction and therefore to achieve the correct micrometer dial reading, the .039" will have to be taken away from the full scale reading of the dial.

The tooth can, of course, be cut in just two passes, one down either side of the tooth, but it is recommended that three passes be used. The advantage is that if a mistake is made in any of the offsets it is apparent, as only a small amount of material is removed on the last two cuts, but a faulty setting when using only two cuts may not be so noticeable.

A direct division dividing head may be used but if so then it must be capable of providing a number of divisions equal to four times the number of teeth being, cut. This will allow for the quarter of the tooth setting needed for the blank roll movements of the second and third cuts.

CHAPTER 12

MAKING GEAR CUTTERS

Without doubt the greatest problem the amateur gear cutter has to face is the cost of the gear cutters. There are eight cutters in a complete set and as a set is needed for each DP, CP or module pitch, the number of cutters needed to cover all eventualities is, to say the least, considerable. The cost of such a large number of cutters is far beyond the budget available to the vast majority of hobby workshops but, fortunately, when gears are required for a project they are often all the same DP and if very small pinions can be avoided then maybe two or three cutters will be adequate to cut all the gears required. Gear cutters, owing to their specialized shape, are expensive and purchasing even a few may be considered impractical so the answer is to make your own! The satisfaction of producing a train of gears oneself is very rewarding and even more so if the cutters are also home-made.

In order to understand the problems involved in producing one's own cutters refer to Fig. 88, where it will be seen that the difficulties fall into two distinct elements; firstly to produce the actual profile of the cutter, which must be the same shape as the space between two adjacent gear teeth, and secondly to back off the tooth and so provide the necessary clearance around the cutting edge.

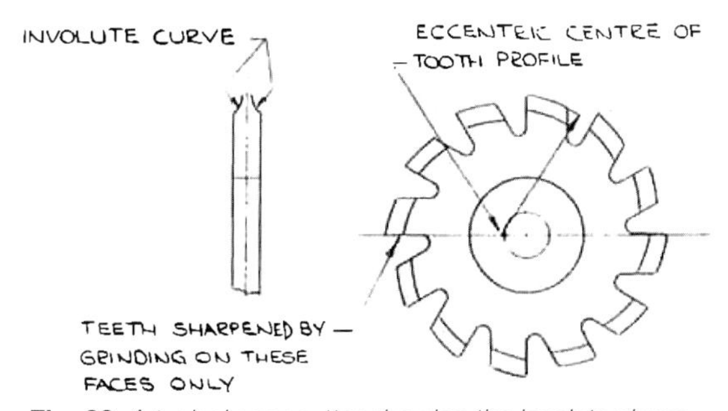

Fig. 88. *A typical gear cutter showing the involute shape and form relief on each tooth*

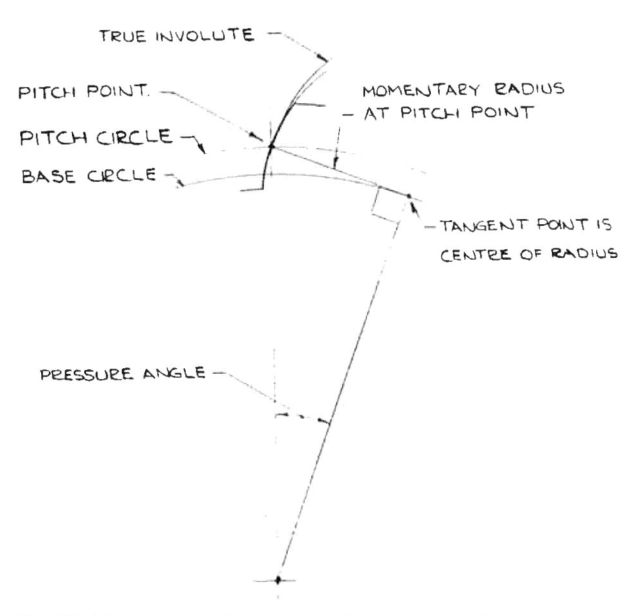

Fig. 89. *How to determine an approximate arc for a button type cutter*

Dealing first with the tooth profile, it is symmetrical about the center line and each side consists of a curve, this curve being the correct involute for the tooth profile required. Referring back to Fig. 24 in Chapter 2, it can be seen that the involute curve is not one that can be readily produced by normal machining methods as it is not an arc struck from any specific center, indeed its "radius" is continually changing, starting with a pronounced curve and then flattening out to follow a rough annular path around its base circle. It can also be seen that only a very small section, right at the start of the curve, is used to form the tooth profile. This small section of the curve can in fact be substituted by a true radius with very little detriment to the gear shape, provided that the correct radius is chosen and that its center is correctly located. The actual error between the radius and the true involute is very small, in fact much less than the error readily accepted by using one cutter only to produce a range of teeth. Being able to substitute a true radius for an involute makes it possible for the amateur to make cutters that are capable of cutting gears to the standard normally accepted in the absence of gear hobbing or other generating machinery.

The method showing how to obtain the size of the true radius, and also how to locate its center, is shown in Fig. 89. The diagram shows that at the "pitch point" both the true involute curve and the substituting radius are coincident and that the actual error in curvature at the outer face of the tooth is very small indeed. Since both sides of a gear tooth are similar in profile then the forming tool to make the gear cutter can consist of two circular hardened cutter bits or buttons of appropriate diameter mounted at some predetermined center distance on a soft steel holder. Such a cutter would produce two

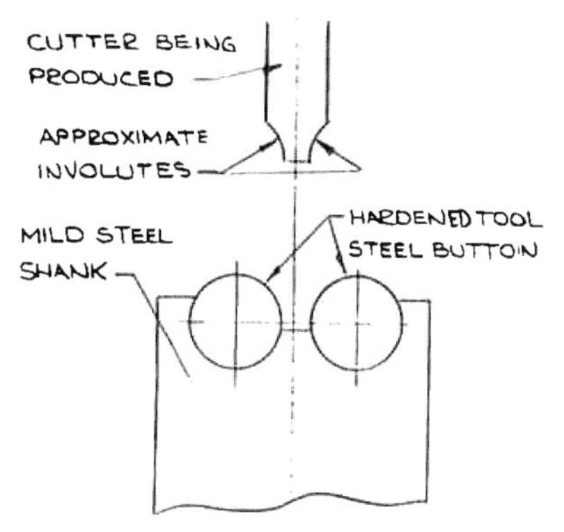

Fig. 90. *Showing the principle of the button tool*

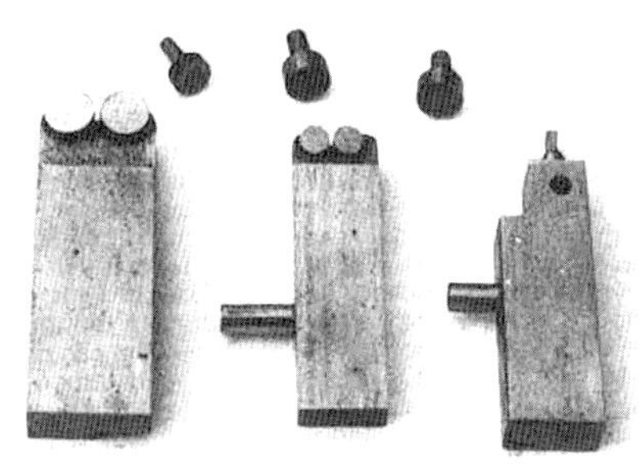

Fig. 91. *This shows two button cutters and a "topping tool." The cutters with the pins in their shanks are for use with the "Eureka" form relieving device. The buttons in the top of the photograph have been hardened and are ready for grinding*

similar segments of circles on a cutter blank and these segments of circles, when applied to the gear blank, would produce the desired curved surface of the tooth. The diagram Fig. 90 and the photograph Fig. 91 show such tools and how they produce the form required on the gear cutter.

The following table lists the button sizes, the center distance between the buttons and also the infeed required for six cutters covering the range from a 17-tooth pinion upwards to a rack. This table is for a pressure angle of 20° and is for wheels for one diametral pitch. For other pitches simply divide the value given by the corresponding diametral pitch required. The width "W" is a suggested minimum width of the cutter and this of course may vary to suit the material available.

One advantage of producing one's own cutters is that non-standard forms can be employed, and once the principles involved have been understood then cutters of any pressure angle or pitch may be produced. It is not suggested that non-standard forms be deliberately adopted but the option is open should any special contingency arise. For instance, if it is an essential element of a design that the pinions must have as few teeth as possible, then the problem can be eased by increasing the pressure angle. Pinions with as few as 10 or 11 teeth become practical if the pressure angle is increased to 30°; naturally all the mating gears must also have the same angle. The following table, similar in concept to the previous one, may be of assistance in producing 30° pressure angle cutters.

Pressure Angle 20°

Cutter No.	Gear Teeth	Dia. "D" In.	Cen. Dis. "C" In.	In. Feed "E" In.	Width "W" In.
1	135-R	51.30	49.60	17.79	4
2	55–134	32.15	31.60	11.47	4
3	35–54	15.07	15.52	5.87	4
4	26–34	10.26	11.03	4.27	4
5	21–25	8.55	9.40	3.71	4
6	17–20	7.80	8.70	3.44	4

The drawing Fig. 92 shows a typical form tool for producing gear cutters. The actual size of the various dimensions will naturally depend upon the range and DP of the gear cutter required but the general method of construction and manufacture will be basically similar for all sizes.

Pressure Angle 30°

Cutter No.	Gear Teeth	Dia. "D" Ins	Cen. Dis. "C" Ins	In Feed "E" Ins	Width "W" Ins
1	135-R	67.50	59.5	17.74	4
2	55–134	27.5	25.00	7.72	4
3	35–54	17.5	16.30	5.20	4
4	26–34	13.00	12.40	4.08	4
5	21–25	10.50	10.25	3.44	4
6	17–20	8.50	8.50	2.95	4
7	14–16	7.00	7.20	2.57	4
8	12&13	6.00	6.36	2.30	4
9	10&11	5.00	5.60	2.11	4

The shank of the form cutter can be made from ordinary mild steel bar section, the actual section depending upon the size and type of the toolpost fitted to the lathe and also on the center distance between the two hardened steel cutting buttons. The shank must be wide enough to provide adequate support for the buttons, as the cutting load upon them is high owing to the long length of cut.

After cutting the shank from the bar material the first operation is to produce the angled face upon which the buttons are mounted. All cutting tools need rake and clearance angles around the actual cutting edge. In the case of the buttons these angles could be provided by turning them to a taper, giving them a part-conical section. Although this may appear to be the simplest way of providing the clearance, it has its drawbacks as any metal removed from the top surfaces of the buttons for sharpening purposes would naturally reduce the diameter and the correct form would be lost. It is far better to use a parallel button and mount it into the shank at an angle. The actual angle is not critical, between 4° and 5° will be satisfactory.

Ideally, the milling of the angle-face and the drilling of the button mounting holes should all be done at the same setting in order to guarantee the holes being square to the face. The vertical milling machine is ideal for this operation as a high degree of accuracy can be achieved without the need for any additional measuring equipment. Grip the cutter shank in a machine vise at the required angle of 5°;

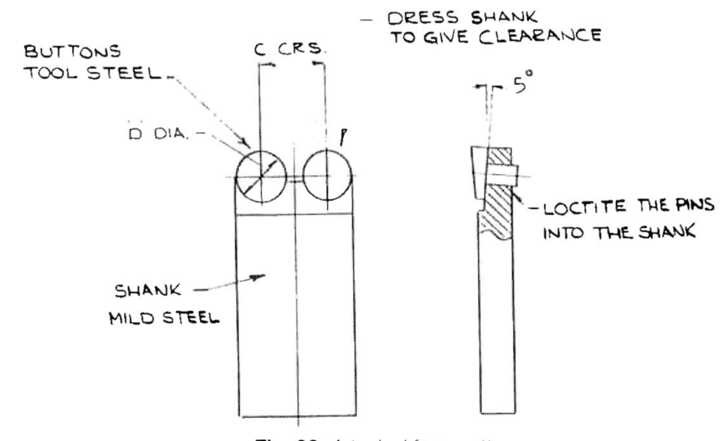

Fig. 92. *A typical form cutter*

a protractor may be used but the angle can be obtained quite easily and more accurately by using the machine itself for measuring the angular setting. Place a lathe center or similar pointed device in the mandrel of the machine and bring it down onto the "high" end of the shank then set the micrometer dial to zero. Move the table two inches—this will be 20 turns on a .100" pitch lead screw—and then, by means of the quill fine feed, lower the marker until it once again just touches the cutter shank. When the difference between the dial readings is .175" then the shank will be set at an angle of 5° to the milling machine table.

With either an endmill or a flycutter, face sufficient from the shank to provide a seat for the button. It may be necessary to actually form a recess on the shank, otherwise when the buttons are placed in position they may be too high and unable to be set at the lathe center height. This point needs to be checked before any cutting takes place. The drawing shows a recess because the author uses a Dixon-type toolholder in his Myford lathe and the recess becomes necessary when using shanks made from ⅜" thick material.

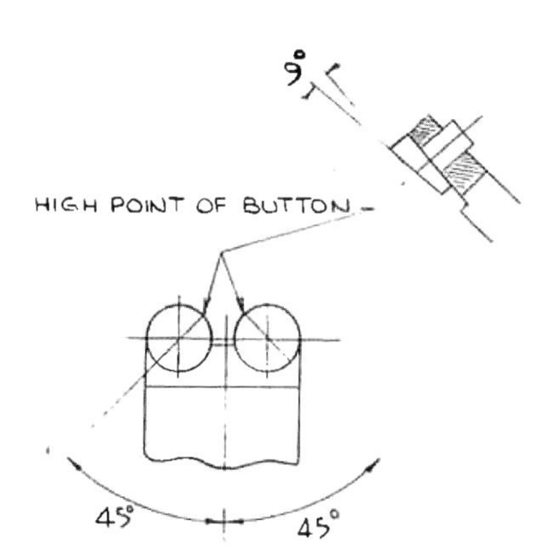

Fig. 93. *Showing how the buttons are positioned in the shanks*

The holes for the stems of the buttons should be drilled without disturbing the setting of the shank, as this will then ensure that the holes are square with the angled face. For many button sizes a stem diameter of ¼" will be adequate, but this may have to be varied for the very small and the large buttons. The distance between the two holes is obtained from the chart and will be the relevant "C" reading divided by the DP being used. Once again, there is no need to measure with any special measuring tools; position the first hole in a convenient place near the edge of the shank and then obtain the distance between the two holes by means of the table feedscrew and micrometer dial.

The buttons will have to be made from tool steel; drill rod is ideal as it is readily available in bar form, relatively easy to machine and simple to harden and temper. Turn the stems down to an easy but not sloppy fit in the holes in the shank, the main diameter "D" in the chart being obtained in a similar way to the "C" center distance. Part off, leaving the head about 5⁄32" or so thick.

The front clearance angle of the cutting buttons has been taken care of by drilling the mounting holes at an angle of 5° but a top rake will also be required—about 3° or 4° will be satisfactory. Owing to the inclination of the pins needed to obtain the clearance angle, 5° top rake will be required just to offset this angle and so to obtain the actual 4° top rake the total angle of the top face of the button will be 9°. This can be achieved either by hand filing or milling in a suitable jig. Anyone

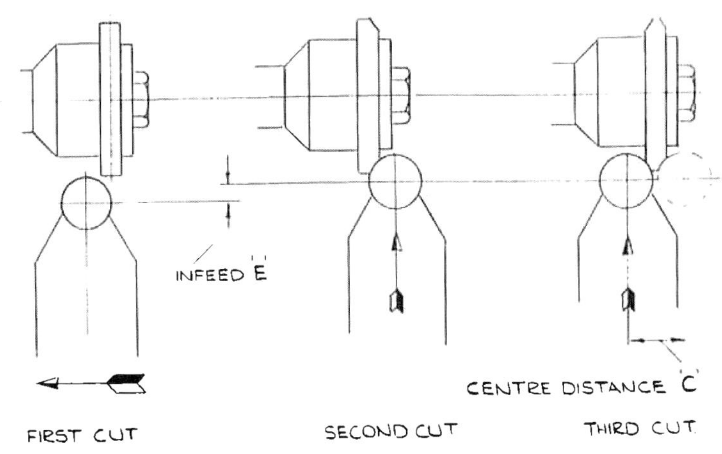

Fig. 94. *Forming both sides of a gear cutter using a single button form tool*

possessing one of the tool and cutter grinders designed for the home workshop, such as the Quorn or Kennett, will be able to grind the angle after the buttons have been hardened. If a suitable grinder is not available then the final finishing will have to be done by careful manipulation of an oilstone slip. Hardening is done in the usual way of heating to a cherry red and quenching in water or oil, depending on the drill rod used, and tempering to a light straw color.

The buttons are secured into the shank by means of Loctite, which will allow them to be moved to the correct angle after they have been placed into their respective holes. The "high" edges of the buttons need to be at an angle of about 45° to the center line so as to derive full benefit from the rake and clearance angles, see Fig. 93. The tool is now ready for use.

Both sides of a gear cutter are the same profile and so both buttons are the same diameter. It is possible, therefore, to produce a forming tool with only one button and procure the correct spacing of the two curves by a measured displacement, which can readily be obtained by manipulating the lathe top slide or carriage. The one-button cutter is of course simpler and quicker to make but it does call for more care in its use, and the displacement between the cuts has to be performed every time a gear cutter is made. The two-button type is simple to use and the chance of an error is minimized. It is simply a case of swings and roundabouts. The diagram Fig. 94 illustrates the use of a single-button cutter.

Now that a means of producing a practical alternative to the true involute has been solved, the remaining problem is how to provide a suitable "back-off," or relief to the teeth of the cutter. Reference to Fig. 88 will show that the whole tooth shape is not produced concentric with the cutter center, since to do so would result in a tooth with no clearance or rake angle and as a consequence it would not cut but merely rub. The problem

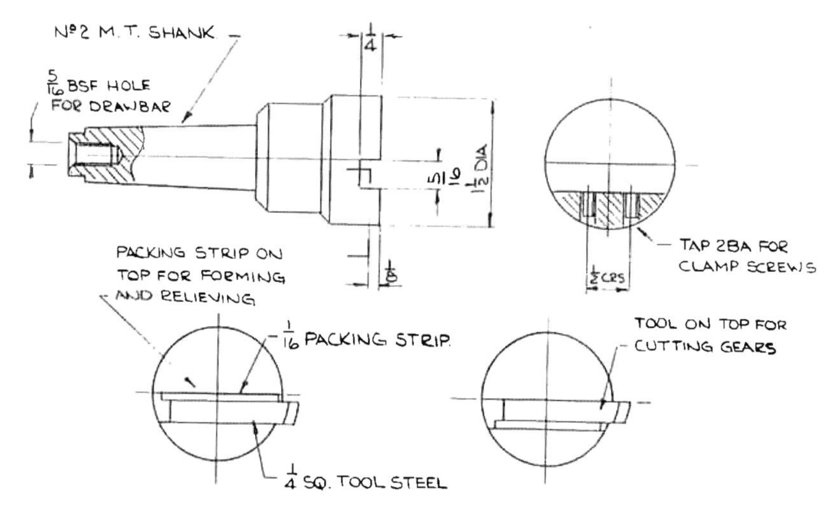

Fig. 95. *Arbor for both producing and mounting a gear tooth flycutter*

of providing a relief to all form cutters is increased by the fact that the shape of the tooth must not be affected when it has to be ground in order to maintain a keen cutting edge. To alter the shape of the tooth in any way would of course affect the form it would produce and so destroy the whole concept of the cutter.

The method used to apply relief to normal side and face milling cutters cannot be adopted as this type of cutter always loses a little of its width during the sharpening process. The solution is to apply curved relief, the whole tooth being produced about an eccentric center, Fig. 88, which maintains a constant cross-section throughout the length of the tooth and because of the tooth eccentricity to the arc of rotation, all its surfaces "fall away" from the front or cutting face and so no side clearance is required. When sharpening, so long as the grinding is limited to the front face, the form will remain constant and the cutter can be ground many times until the tooth finally breaks away owing to weakness.

It is neither an easy nor straightforward machining operation to apply curved relief to a multi-tooth circular cutter; commercially made cutters are "backed-off" on a special form-relieving lathe and this time-consuming process is the main factor in the high cost of gear cutters. Turning plain eccentric shafts is, however, a simple operation to carry out in the home workshop. After being machined in its normal centers, the workpiece is either mounted in offset centers or moved over in a four-jaw chuck so that the eccentric element can be produced by normal turning operations. This is the way model engineers make their valve gear eccentrics for steam engine models.

If the cutter, Fig. 88, is first turned and formed about its normal center, teeth can then be produced by milling a series of gashes into the cutter disc. However, these teeth—being concentric with the cutter axis—will have no relief and so will not cut, but if we mount the cutter on an eccentric mandrel we can eccentrically turn one tooth about the new center and so produce a correctly relieved tooth form. There is one big snag in doing this in that every tooth has to be formed about a different eccentric center and so the act of relieving one tooth will have totally destroyed many of the teeth that follow it. In fact, only one tooth out of the complete disc can be successfully relieved by normal eccentric turning methods. If we cut away from the disc all the mutilated and unaffected teeth we are left with a single-tooth cutter, commonly termed a fly-cutter.

The fly-cutter is a very useful type of cutter and if suitably held will perform all the duties of a normal circular cutter. The main difference is that as only one tooth is cutting for each revolution of the cutter, the swarf removal rate is naturally lower and the overall production time increased but not, surprisingly, by a great amount! Single-pointed fly-cutters are perfectly satisfactory for gear cutting, in fact they have one advantage over the multi-tooth cutter in that they are extremely simple to keep sharp.

There is an old fable that says roast pork was discovered when a pigsty was burnt down and when the farmer, while removing a hot carcass, burnt his fingers. He put his fingers into his mouth to ease the pain and discovered that the hot substance on his fingers had a pleasant taste. The farmer soon learned that there was no need to burn down a sty every time he wanted roast pork, and it is the same with a single-tooth form cutter. A fly-cutter can be made without producing a disc and then cutting a strip off it. A piece of square section tool steel is all that is required and, once again, tool steel such as O1 or W1 (the same as drill rod) is the obvious choice. The size of the tool will naturally depend upon the tooth size it is going to produce: for 20DP a ¼" square section is recommended, but for small tooth forms then ³⁄₁₆" or ⅛" square material may be used.

In order to make the cutter, some form of holder will be needed in which to mount it while the forming and backing-off operations are undertaken. A holder will also be required to support and drive the tool during its cutting operations and with a little ingenuity one holder can be adapted to perform both duties. It can be used to support the cutter while it is formed then, after the cutter has been hardened and ground, the tool can be mounted in the same holder for the actual gear cutting process.

Fig. 95 shows such a holder, the key to its versatility being the little packing piece. This is used in the holder both for the forming of the cutter and also in the general use of the finished tool. The slot in the end-face of the holder is made wide enough to accommodate both the cutter and also the packing piece. The slot is positioned so that its top surface is on the center line of the holder. For the forming operation the packing strip is placed above the cutter, this then positions the top face of the cutter below the center line of rotation. Now, by ordinary turning methods, and with the holder rotating truly about its own axis, the button tool will correctly form and at the same time "back-off" the cutter relative to its working face. After hardening and grinding, the cutter is placed back into the holder but this time the packing piece is positioned below the cutter, thus bringing the top face of

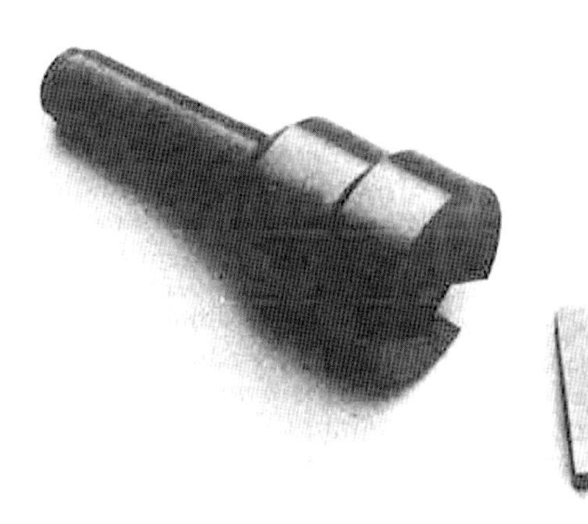

Fig. 96. *The holder, flycutter and packing strip detailed in Fig. 95*

the cutter coincident with the center line of the holder and into its correct operating position.

To sharpen the cutter the whole of the top face is ground and this operation can be performed many times because the thinning of the cutter has no effect on the operating attitude as its top, or cutting, face is always hard against the upper face of the slot and so on the center of rotation.

In eccentric turning the normal center line of the cutter is displaced by a certain amount relative to the tooth being cut. In the case of the holder and packing strip, the center line is unaffected but the tool is offset; this is a simpler process but the end result will still be the same.

Another method of producing a fly-cutter is shown in Fig. 97, the cutter in this case being a circular one with the tooth concentric to the bore. No eccentric turning or "backing-off" is needed so the tooth profile can be machined at one setting on a true-running mandrel, the button tool again being used to produce the appropriate profile. The form relief is achieved by mounting the cutter eccentrically onto its operating arbor, which unfortunately in this case is not suitable for the initial turning. This type of cutter actually has two teeth but only one can be used at a time, the advantage being that if it becomes necessary to sharpen the cutter during actual gear cutting then the job need not be stopped as the cutter can be turned over and the second tooth brought into use without disturbing any part of the machining setup. When not being used for cutting, the second tooth still has a duty to perform as, by means of a pin, it provides a positive drive to the cutter. This will eliminate any tendency the cutter may have to slip under the load which, if this occurred, would reduce the eccentricity and destroy the depth setting.

The arbors for both this and the previous tools are shown with Morse taper shanks for direct fitting into the milling machine or lathe spindle. This type of mounting is superior to chuck mounted arbors as they will always run truly, which is essential for a gear cutter, and

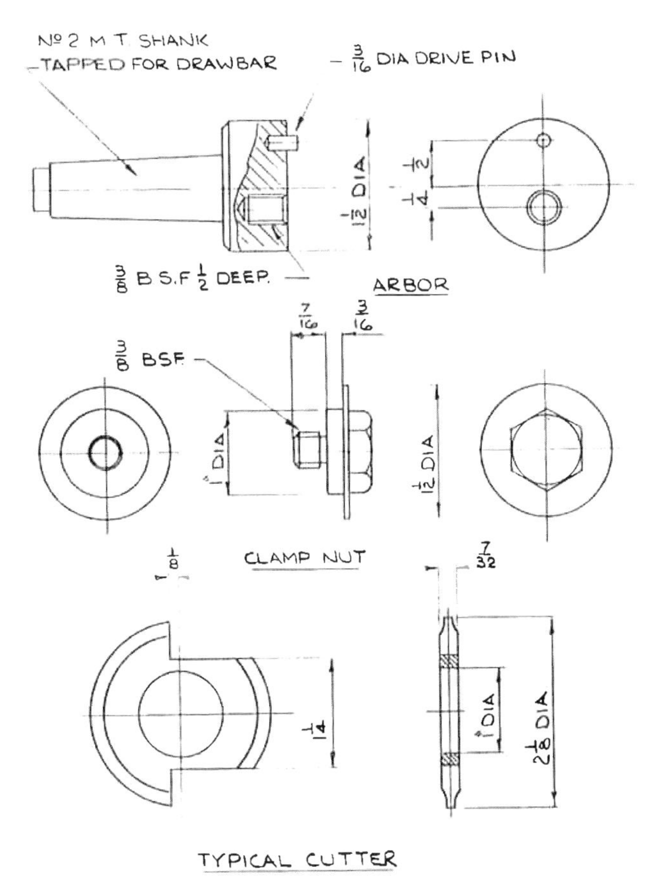

Fig. 97. *Arbor for circular flycutter; the sizes may vary to suit design requirements*

they also provide greater rigidity to the cutter. Another advantage is that the machine can be set to rotate in either direction without the fear of the chuck unscrewing from the mandrel.

The type of cutter shown in Fig. 100 is not provided with a driving pin, relying solely on its clamping nut to secure it to the driving arbor. It is intended for small cutters where the cutting loads are lower. It is a circular form cutter with a cut-out portion to provide it with one cutting tooth and the relief is provided by boring its mounting hole eccentrically from the normal axis, the center of the hole being positioned diametrically opposite the cut-away portion.

Although this is basically a simple type of cutter its manufacture does present a small problem because the outside profile and the bore cannot be machined at the same setting. There are one or two ways in which the problem can be overcome and one is to produce the cutter from bar material. Making cutters from bar has the advantage that most of the machining can be carried out at one setting and then parted-off to the required thickness. After parting-off, the hole can be produced either by drilling in the normal drill press or by setting eccentrically in the four-jaw chuck and producing the bore in the lathe.

An entirely different approach is to make the cutters from plate material such as gauge plate; this is a good material for cutters as it is easy to machine and to harden. Being specially made for the manufacture of gauges the amount of distortion suffered in the hardening process is virtually nil. To produce a blank, saw out from the sheet an approximate disc a little above the finished diameter of the cutter and drill a hole in the center of it. The diameter of the hole must be no more than a half of the finished bore; for instance, if the cutter is eventually to be mounted on a ⅜" dia.

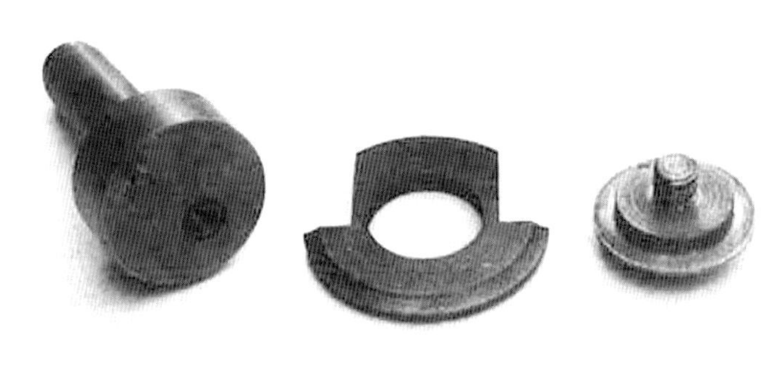

Fig. 98. *The cutter and arbor shown in Fig. 97*

Fig. 99. *The two flycutters assembled and ready for use*

spigot then the initial, or service, hole should be no more than 3⁄16" dia.

The actual diameter and profiling with the button tool is carried out on a mandrel but make sure that the flat clamping face of the mandrel is square with the spigot diameter. The end of the spigot is screwed for the provision of a clamping nut. Care will be needed, particularly with the forming operation, as the cutting forces will be relatively high and the spigot diameter small, but a cautious approach will ensure a satisfactory result. After forming, the small concentric hole will have to be made into a large eccentric one and this will have to be done by setting up the blank in a four-jaw and using a single-point boring tool. The minimum diameter of the finished bore cannot be less than the diameter of the original service bore plus twice the amount of eccentricity.

With this type of cutter the eccentric element is in the cutter itself, and if under load the cutter slips on the operating mandrel it will not affect the setting of the cutter in any way as the eccentricity providing the relief is constant for all positions of cutter and arbor.

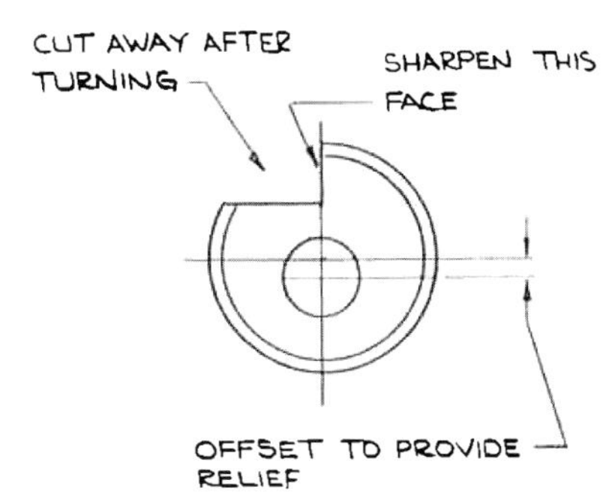

Fig. 100. *A cutter suitable for producing small size teeth*

MULTI-TOOTH CIRCULAR CUTTERS

The single-point fly-cutters are perfectly satisfactory for cutting the "one off" type of gear but if it is intended to cut gears on a regular basis then it will be well worthwhile building up a stock of multi-tooth circular cutters. This type of cutter can be produced successfully in the home workshop but it does necessitate the making of some special tooling equipment. As already outlined, the main problem in making multi-tooth cutters is to provide the correct form relief to each tooth. Over the years many attachments have been designed and made to perform this operation and most of them work very well, the main criticism being that they only deal with one tooth at a time and so the complete process of form relieving a cutter of twelve teeth or more can be somewhat tiresome.

As it happened, Professor Chaddock and the author had been interested in this problem for some time when a copy of a page from an old Victorian tool catalog advertising a continuous form relieving attachment came into their possession. It was sent by a reader of *Model Engineer* magazine and curiosity was immediately aroused, although, unfortunately, the advertisement gave no details as to how the device worked or how it was constructed and a search through old books of the period proved fruitless. Nevertheless, this old print gave the necessary inspiration and after some discussion a tool was designed and made which performs very well. It was given the name "Eureka" and a full detailed description of it has been published in *Model Engineer*.

The tool is used mounted between the centers of a lathe and, after giving the correct backing-off action to a tooth on a gear cutter,

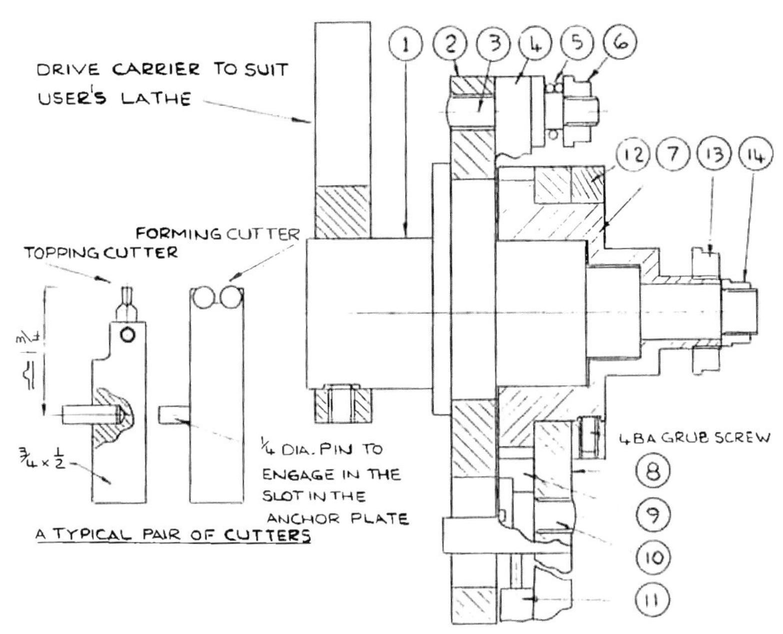

Fig. 101.

it then indexes it around to the next tooth and so produces a continuous relieving action. The tool will only relieve a cutter that possesses the same number of teeth as its ratchet wheel, but as there is no detriment in making all cutters with the same number of teeth this limitation is of no consequence. The device can be made to any size within reason but the prototype, and the one shown on the drawings, is for cutters of a half-inch bore and possessing twelve teeth. It will successfully relieve cutters up to about 20DP and of 1¼" to 1½" diameter.

This size was chosen because material for making the cutter, such as drill rod or gauge plate, is readily available and can be heat treated without specialized equipment. This size of cutter is adequate for most model engineers' general needs but anyone wishing to cut gears with a larger tooth profile would need to increase the size of the relieving tool. The number of teeth on the ratchet wheel may also be increased but even with a cutter of a larger diameter it is doubtful if any advantage would be gained by the addition of more teeth.

The Eureka device is neither difficult nor complicated to make—the prototype was completed and in operation within four days of the design being finalized. It is not so easy, particularly from a study of the drawings, to visualize the cycle of events but once the tool is seen in operation then its mystery quickly unfolds.

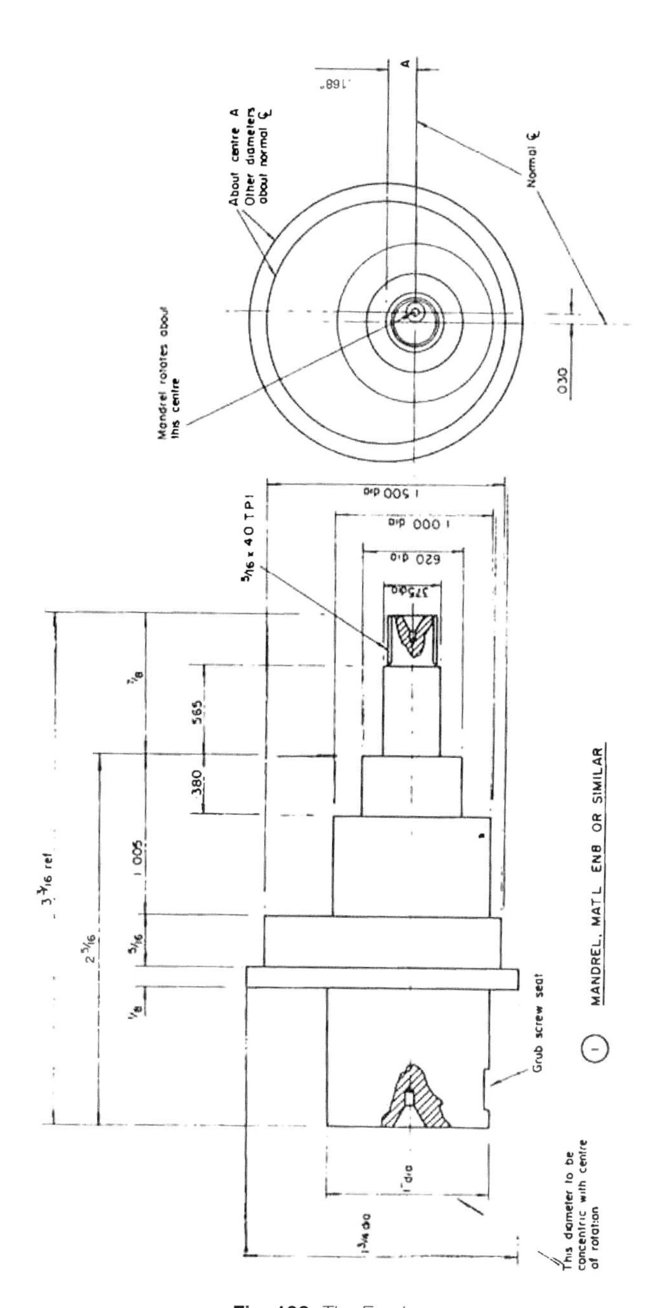

Fig. 102. *The Eureka*

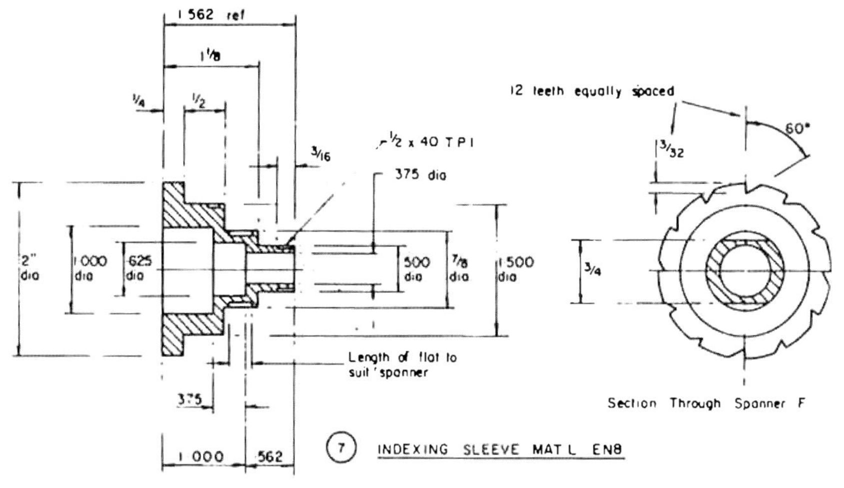

Fig. 103.

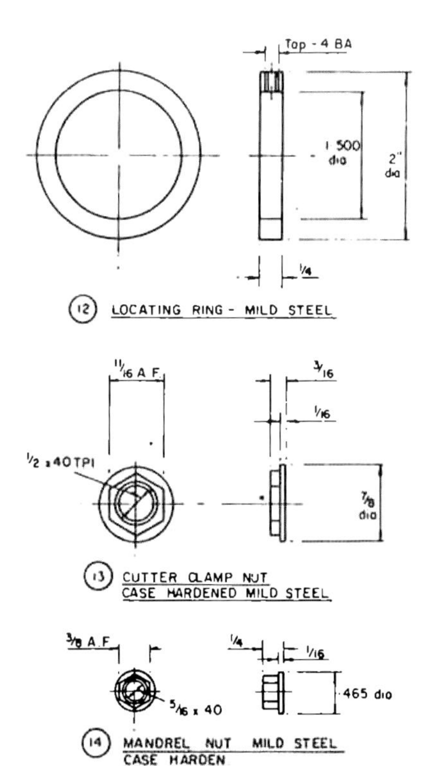

Fig. 104.

The device is an eccentric turning attachment operated by two eccentrics on the same shaft or mandrel. When placed in the lathe between two centers and rotated continuously it will, by means of one eccentric, index each tooth in turn while the other—and smaller—eccentric provides the movement needed to give the tooth its form relief. The first, or indexing, movement is not difficult to visualize; the rearmost plate, item 2 on the drawings, is the ratchet plate and is mounted on the largest of the eccentrics. It is prevented from rotating by an arm engaging in a fixed pin secured in the anchor plate (item 8), which is itself prevented from turning by the pin in the cutter holder. The ratchet plate will therefore oscillate with respect to the ratchet wheel, one pawl engaging in the teeth of the ratchet and alternately gathering and indexing the ratchet wheel.

The second pawl, secured to the anchor plate, acts as a back stop and prevents reverse rotation during the gathering phase.

We now have a means of indexing the blank correctly provided that both ratchet wheel and gear blank have the same number of teeth. As it stands the blank would merely rotate about a fixed center and no relief would be obtained so it is the duty of the second eccentric to provide this relief action. The cutter seating, being made to rotate eccentrically about the normal axis, advances and retracts the cutter blank to and from the fixed form tool. The timing between the two movements is important and is outlined below.

As the ratchet pawl is gathering, the ratchet wheel (item 7) and the cutter that is attached to it are not rotating, although they are moving under the influence of the relieving eccentric which, when the gathering commences, will start to move the cutter away from the form tool. However, before the gathering is completed the relieving eccentric will have started to move the cutter blank back towards the form tool so that when the gathering stops and the indexing and cutting action begins, the cutter blank will have advanced up to the form or button tool and cutting will begin. The cutter blank will be moving towards the form tool and providing the form relief the whole of the time that it is rotating. When the tooth being cut has completely passed the form tool and the gap between two teeth has been reached the indexing will stop and the cutter blank will stop rotating. At the same

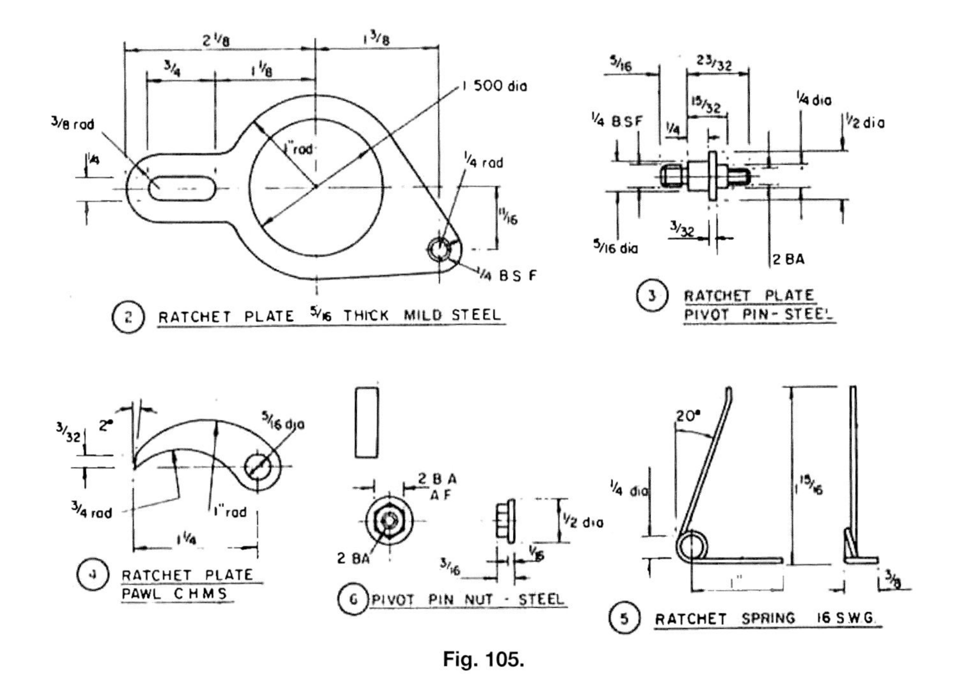

Fig. 105.

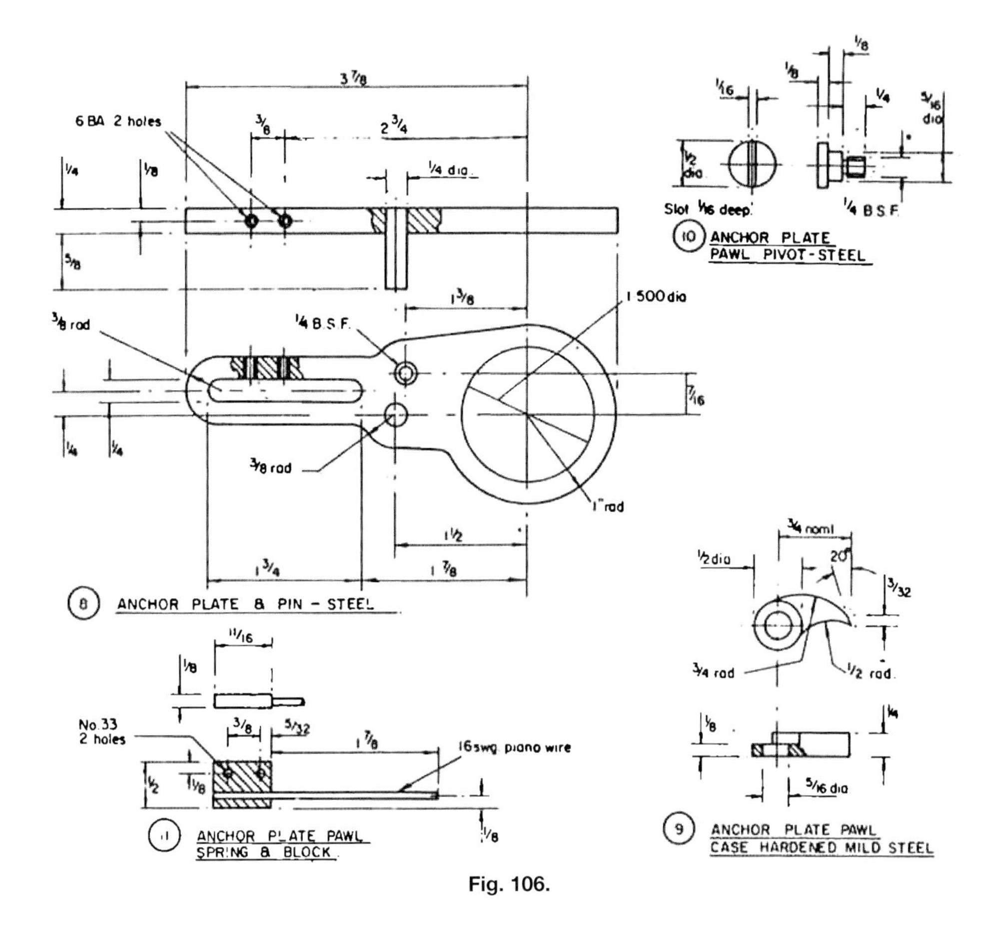

Fig. 106.

time the relieving eccentric will have passed its full extent of travel and will have started to withdraw the cutter away from the form tool ready for the next cycle of events which commences with the indexing of the next tooth.

It is not intended to give a detailed account of the various machining operations needed to build the tool as this is not the place to discuss general machine shop practices and techniques, but a full and detailed set of drawings is shown in the Figs. 101 to 107 and provided these are carefully followed no problems should be encountered. Care should be taken to ensure that the timing of the two eccentrics on the main mandrel is correct as an error here could prove disastrous to the correct functioning of the device. No tolerances have been quoted on the drawings, the reason being that it is the fit between components that is important rather than their basic sizes. Try to achieve a close but not tight

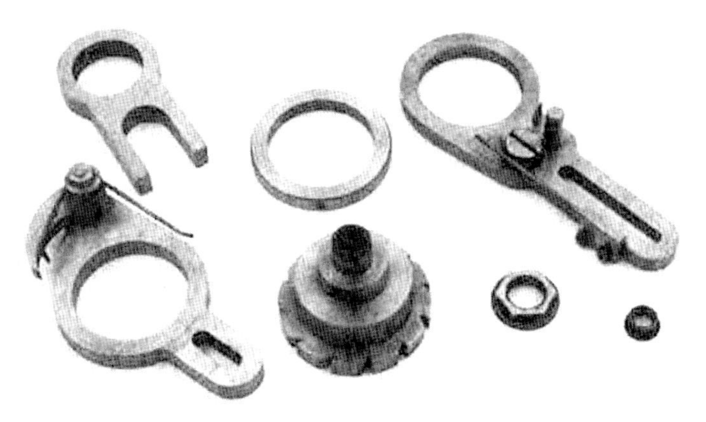

Fig. 107. *Above, the spread of parts needed to make the Eureka form-relieving tool. Right, the completed machine with a cutter mounted. Below, in use on the lathe*

fit between the 1.000" and .375" diameters on the mandrel and their respective bores in the index sleeve (item 7). The Eureka is not a difficult tool to make and the average model engineer will have no problems in quickly constructing this rather unique tool.

Once the necessary tooling has been made the manufacture of the actual gear cutter can be considered. As previously stated the form relieving tool is intended to relieve cutters of about 1¼" dia. and Fig. 108 shows a typical cutter of this size. The first stage is to produce the blank and this can be made from either bar or plate material. If bar material has been chosen then this should be set to run truly in a four-jaw chuck and the end faced. The bore can be roughed out by means of a drill, or series of drills, but it is recommended that a single-point boring tool is used to bring the hole up to size. If a half-inch reamer is available then this can be used to finish and size the bore but only allow about .005" or so for the reaming process.

The next operation is to part-off to thickness and this operation can be a little tricky, particularly if the lathe being used is on the flimsy side—a tail support can be used to advantage here even if the overhang is small. To obtain a good surface finish on the side face, start to part-off a little on the thick side, do not complete the cutting off but stop leaving a wall thickness between the cut and the bore of about 1⁄16". Then, withdraw the cutter and reset to the required thickness and slowly feed the parting tool in again, this time using a little cutting oil and part-off completely. Once the lathe has been set up for this operation it is as well to produce a number of blanks even if they are not all initially required as they can be put aside for future use and so save time at a later date.

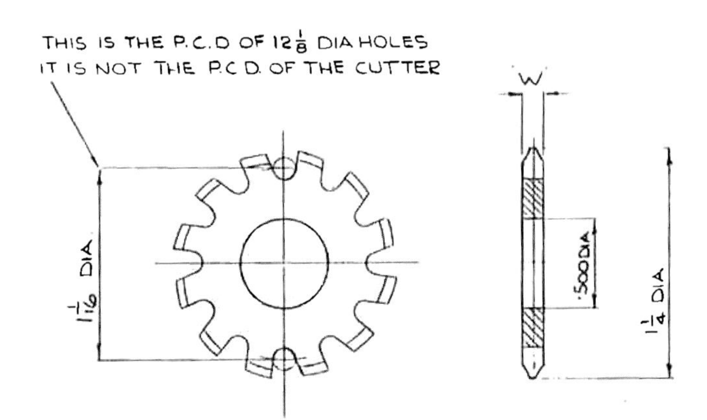

Fig. 108. *A typical cutter blank suitable for the Eureka relieving device*

Fig. 109. *Drilling the holes in a cutter blank. The indexing is obtained by means of a rotary table*

If the blanks are to be produced from sheet material such as gauge plate then roughly cut them to profile before machining the hole. The bore must be square to the face and in order to achieve this it may be advantageous to clamp the blank down flat against the face-plate rather than trying to set it up in a chuck. To turn the outside diameter, mount the blank on a mandrel as this will ensure concentricity between the two diameters.

The next operation is to drill the twelve equally spaced holes that form the basis of the gaps between the teeth. A simple drilling jig can be made to position the holes correctly but if a rotary device is available then this can be used to advantage. The photograph Fig. 109 shows such a device in use. The center spindle of the device is set directly over the spindle axis of the milling or drilling machine and then by using the table feedscrew and micrometer dial the correct offset or radius for the holes can be obtained. The indexing between each hole is then achieved by means of the handwheel on the rotary table.

The actual gap between the teeth can be produced by carefully filing down and into the drilled holes but an easier method is to remove the unwanted metal by means of a slitting saw mounted on a suitable arbor in the spindle of a vertical milling machine. The cutter blank, still on the arbor that secured it for the drilling operation, is then placed horizontally in the dividing head. Two cuts are needed to form each gap, both cuts being parallel to each other, and the only movement needed between each cut is a lowering of the slitting saw. This operation should be performed on all twelve teeth before setting for the second cut. The illustration Fig. 110 shows the principle of the operation in diagrammatic form, while the photograph Fig. 111 shows the actual cutting being performed and the tooling used.

The cutter is now ready for profiling, this being done by means of the button tool. The cutter is mounted on an arbor and once again it is the same arbor used for the previous operations but this time it is secured in the lathe spindle. The depth of the cut, or the amount the button tool is advanced, is of vital importance for it is this that will determine the final shape of the gear teeth. The distance to advance the buttons is as shown in the chart and under the heading "In Feed—'E' Ins" (see pages 112-113). Read off the figure appropriate to the cutter being made and then divide this figure by the DP required. The result will be the required depth of cut. The depth should be obtained by means of the cross-slide micrometer dial. First set the extreme front edge of the button so that it just touches the outside diameter of the teeth on the cutter then set the micrometer dial to read zero and centralize the buttons on the cutter. Run the lathe in a low back gear speed and carefully and slowly feed in the button cutter until the required depth has been reached.

The last machining operation is to apply the form relief, which is done with the same button tool used for the initial profiling. This tool now actually becomes part of the Eureka device as mounted in the shank of the tool is the pin that prevents the ratchet and anchor plate from rotating. The angular position between the cutter to be relieved and the device is important and must be correctly set otherwise the relieving sequence will be mis-timed. The indexing or rotary motion of the cutter must start in one of the gaps between the teeth and cease in the following gap.

This setting is best obtained by trial, so put the cutter in the device but only grip it lightly so that it may be rotated on its seating by hand. Then advance the button tool almost up to the cutter and rotate the lathe mandrel; it will be apparent at once whether or not the angular setting is correct. If not, adjust the position of the cutter and try again. Once the setting has been determined tighten the clamp nut securely using two spanners, one on the nut and the other on the ratchet body. If this second spanner is not used the whole of the tightening force is taken on the ratchet pawls and this is not desirable.

The actual cutting operation can now begin. Start by advancing the tool by small increments of about one-thousandth or so and do not forget to allow the cutter to make one complete revolution, or twelve cutting events, before advancing the tool a further increment. When the cutting starts it will remove metal from the back, or heel, of the tooth only but gradually the cut will lengthen. When the cut reaches the leading edge of the tooth then this part of the form relieving operation is complete. The buttons will only have given relief to the sides, or circular arcs, of the teeth and not to the outer lips, or tops, as these will have passed between the two buttons and so remain unaffected. A change of tool will therefore be needed, the second tool being merely a square-ended tool similar to a parting tool. The width of the tool tip is not important but it must obviously be wider than the tips of the teeth to be cut. This second operation is a simplified version of the initial process: carry on cutting until the length of the cut just reaches the front edge of the tooth. The cutter is now completely relieved and is ready for heat treatment.

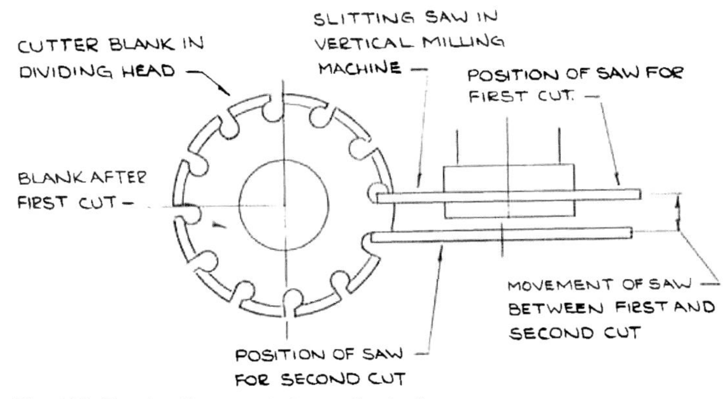

Fig. 110. *Forming the gaps between the teeth*

Fig. 111. *Showing the gullets being formed by means of a dividing head and slitting saw*

The hardening process will depend upon the material used but in the case of drill rod, heat to a cherry red and then plunge it into water for W1 steel, or oil for O1. It will be necessary to temper the cutter, otherwise the teeth will be brittle—but do not over temper, heat until the color just begins to change and then quench again. Grinding the faces of the teeth will be a relatively simple task for owners of a cutter grinder such as a Quorn or Kennett, but if one of these machines is not available then carefully dress the face of each tooth with an oilstone slip and the cutter is ready for use.

The author hopes that the reader who may at first have been rather hesitant about gears has gathered confidence and now realizes that the cutting of a gear is well within the scope of the average model engineer. A great deal of satisfaction can be achieved from producing one's own gears and, after all, the satisfaction of creating things with one's own hands is the basic driving force behind the amateur engineering movement. If, therefore, this book has encouraged, inspired or enthused a few constructors who previously regarded gear cutting as being outside their ambit, to "have a go" then the task of writing it has been well worth while.

APPENDIX

The tables shown on pages 112-113 are tabulated for 1DP. All the dimensions shown must be divided by the DP required.
For instance, if a 20DP cutter is wanted, then all the numbers shown in columns D, C, E and W must be divided by 20. Only a small proportion of the button's circumference is used for cutting, as can be seen by studying the drawings above the tables. The rest of the button is not used and can be discarded. When the buttons are of a small diameter (similar to those shown in Fig. 91) there is no point in doing this, they may be left as complete discs.

As the button diameters get larger they will naturally overlap. In this case, the overlapping portion of the buttons will have to be removed. This is of no consequence as this part of the button has no duty to perform.

When the DP is a small number and the number of teeth desired to be cut is large, then producing a complete disc for the button may not be practical, so some other means of producing the cutting arc may have to be found, such as a rotary table or in extreme cases marking off and hand filing. The distance between the two arcs "C" in the table is very important and must be maintained no matter how the cutting arcs are produced. It is most unlikely that large gears needing large button diameters will be required by the model engineer and the turned circular buttons detailed in the book will, in the majority of cases, be all that is required.

INDEX

Note: Page numbers in parentheses indicate non-contiguous references.

abbreviations, for equations, 54–55
arc of approach, 20, 54
arc of recess, 20, 54

base circle, 23–25, 54, 62
bevel wheels, 35–44
  crown wheel and pinion, 41–42
  cutting bevel gears, 102–9
  gear teeth, 39–41
  how they work, 35–39
  laying out pair of, 42–44
book (this), overview of, 4

center distance, 8–9, 55, 56
chordal thickness (CT), 54, 55
circular pitch (CP), 27–28, 53, 55
circumferences, calculating, 8
clearance, defined, 54
cones. *See* bevel wheels
constant velocity
  bevel gear teeth and, 39–41
  cycloidal curves for, 14–19, 23
  involute teeth and, 23
  tooth shape and, 14
contact, tooth, 19–20
crown wheel and pinion, 41–42
curves
  cycloidal, 14–19, 23
  epicycloid, 16, 17, 32
  hypocycloid, 17, 18, (32–34)
  involute. *See* involute teeth/form
cutters, making. *See* gear cutters, making
cutting bevel gears, 102–9
cutting spur gears, 62–82
  angle of inclination table, 75
  calculation tables, 63–64
  in horizontal milling machine, 80–82
  methods and guidelines, 62–75
  RPM table, 65
  table of cutters, 63
  in vertical milling machine, 76–80
cutting worms and worm wheels, 83–101
  determining changewheels to produce helix angle, 87–96
  free wheel hobbing, 101
  helix angles, 85–87
  methods and general guidelines, 83–87
  producing worm wheels, 96–100
  tables for, 85–86
cycloidal curves (teeth)
  about, 14–19, 23
  rack and pinions with, 32

D+f, 29, 53, 55, 63
definitions, 53–54
diametral pitch (DP), 28, 54, 55–56. *See also cutting references*
dividing heads, 57–61
drivers
  driven discs and, explained, 7, (8–11)
  explained, 7, 10–11
  lantern pinions and, 20–23
  rack and pinion gears, (30–34)
  worm gears, 45

epicycloid curves, 16, 17, 32
equations, 54–56
  abbreviations for, 55
  circumference, 8
  delineated, 55
  distance between centers (center distance), 8–9, 55, 56
  example using, 55–56
  speed and diameter ratios, 7–8

friction
  tooth forms and, 12, 14, 19–22
  worm gears and, 46, 51–52

gear cutters, making, 110–27
  difficulties in, 110–11
  methods and guidelines, 110–21
  multi-tooth circular cutters, 121–32
gears
  basics of, 6–11
  as discs, transmitting motion/power, 6–7
  drivers explained, 7, 10–11
  how they work, 6–11
  learning about, this book and, 4
  more than two. *See* trains

for non-parallel shafts. *See* bevel wheels
what they are and what they do, 6–7

heads, dividing, 57–61
helix angles, 47, (50–52), 85–87
hob, 50, (98–101)
hobbing, free wheel, 101
hypocycloid curves, 17, 18, (32–34)

idlers, 8–11
involute teeth/form
about, 23–26
involute curve defined, 23
of rack and pinion, 31–32

lantern pinions, 20–23, 33, 34
line of action, 20, 54

module (M), 28–29, 54, 55

outside diameter of gear (OD), equations using, 55, 56

path of contact, defined, 54
PCD/PD/PDM. *See* pitch circle diameter
pin teeth, applied to rack and pinions, 32–34
pinions
crown wheel and, 41–42
cutters for, 112
lantern pinions, 20–23
number of teeth, 55–56
rack and pinion gears, 30–33
worm gears and, 45
pitch
circle, defined, 53
circular (CP), 27–28, 53
diameter, defined, 53
diametral (DP), 28, 54, 55–56. *See also cutting references*
pitch circle diameter (PCD/PD/PDM)
cycloidal curves and, 14–19
defined, 11
equations using, 55, 56, 97–98
involute teeth/form and, 23–26
lantern pinions and, 20–23
pressure angle, 25–26, 31, 54, 83, 112–13
proportions, tooth, 29, 54–56
proportions, worm gear, 49, 52

rack and pinion gears, 30–33
root diameter, defined, 53

size of teeth. *See* tooth sizes
spur gears
defined, 12
rack and pinion gears compared to, 30, 31
worm gears and, 45, 46
spur gears, cutting, 62–82

thickness of tooth on pitch line (T), 54, 55
tooth forms, 12–26
bevel gear teeth, 39–41
cycloidal curves and, 14–19
friction and, 12, 14, 19–22
fundamentals of, 12–14
involute teeth and, 23–26
lantern pinions and, 20–23, 33, 34
projections, slots and, 12–14
resistance and, 12
shape of teeth and, 13, 14
tooth contact and, 19–20
"tooth" defined, 13
tooth sizes, 27–29
circular pitch (CP), 27–28, 53, 55
diametral pitch (DP), 28, 54, 55–56. *See also cutting references*
module (M), 28–29, 54, 55
tooth proportions, 29, 54–56
trains
defined, 8
mechanics of, 10–11
setting out from scratch, 18–19

velocity, constant. *See* constant velocity

worm gears, 45–52. *See also* cutting worms and worm wheels
efficiency of, 51–52
helix angles, 47, (50–52), 85–87
how they work, 45–48
tooth shapes of worm and wheel, 48–52